I0828254

Our Lady of the World's Fair

Our Lady of the World's Fair

Bringing Michelangelo's Pietà *to Queens in 1964*

Ruth D. Nelson

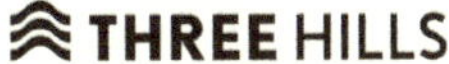
AN IMPRINT OF CORNELL UNIVERSITY PRESS
ITHACA AND LONDON

First published 2024 by Cornell University Press

Library of Congress Cataloging-in-Publication Data

Names: Nelson, Ruth D., 1957– author.

Title: Our lady of the World's Fair : bringing Michelangelo's "Pietà" to Queens in 1964 / Ruth D. Nelson.

Description: Ithaca : Three Hills, an imprint of Cornell University Press, 2024. | Includes bibliographical references and index.

Identifiers: LCCN 2023059473 (print) | LCCN 2023059474 (ebook) | ISBN 9781501776908 (hardcover) | ISBN 9781501776922 (epub) | ISBN 9781501776915 (pdf)

Subjects: LCSH: Michelangelo Buonarroti, 1475–1564. Pietà—Exhibitions. | New York World's Fair (1964–1965 : New York, N.Y.)—History. | New York World's Fair (1964–1965 : New York, N.Y.)—Influence. | Art and society—History—20th century.

Classification: LCC T786 1964. B2 N45 2024 (print) | LCC T786 1964 .B2 (ebook) | DDC 730.92—dc23/eng/20240206

LC record available at https://lccn.loc.gov/2023059473

LC ebook record available at https://lccn.loc.gov/2023059474

Contents

Illustrations

Preface

The 1964–1965 New York World's Fair introduced us to the World of Tomorrow, the world we live in today. There never was nor ever could be a better fair, and that is the memory I have carried since that family vacation brought us to the Queens fairgrounds in 1964. Though I do not remember much, what remains in my heart is a sense of wonder and happiness. It was for me, as it was for a generation of young baby boomers, my introduction to the *Pietà*, held as one of the greatest works in Western art. Though the 1964–1965 fair is best remembered today for its iconic Unisphere, it also marked the first and only time that the *Pietà* was allowed to leave St. Peter's Basilica, and this singular moment helped make the Vatican pavilion the second-most visited pavilion at the fair, after General Motors.

The awe and reverence surrounding the sculpture impressed upon my young mind the weight of the moment. Michelangelo's depiction of Mary's sorrow and pain at the death of her son has resonated through the centuries as the universal expression of a mother's deep, suffering love for her child. Only in retrospect do I now understand the magnitude of what it meant to transport the *Pietà* to America. It would be comparable to shipping, for example, the Statue of Liberty to Rome. But if that were ever to happen, the outcry would come from Americans themselves. In the case of shipping the *Pietà*, protests came from both sides of the Atlantic. The move was a gamble, for sure, but in retrospect, the exposure to this masterpiece led me, in a way, back to academia to study art history. My very first class was Italian Renaissance Sculpture. For all of Robert Moses's bluster, the fair really was a "summer university."

Ever since the opening of the first "world's fair" with London's Crystal Palace in 1851, education has always been a feature, albeit a lesser one, of these expositions. More often, world's fairs have been venues for the latest displays of industry and technology. As President William McKinley presciently remarked in 1901, "Expositions are the time-keepers of progress,"[1] and seldom was this more apt than at Flushing, Queens, with IBM's introduction of the computer ushering in the new digital era.

Increasingly, scholars are investigating world's fairs for their broader significance as diversions from national and global issues, resulting in a more clinical and even jaundiced analysis. In the case of the 1964–1965 fair, this scrutiny has relied too heavily on the negative coverage by the *New York Times*. I had to ask myself: Did we visit the same fair? Although I, too, was disappointed to learn that the fair was a means to an end, and that was to finance the completion of the Flushing Meadows–Corona Park. How could all that wonderfulness have been accidental? Teaching a class on the fair, I will occasionally meet a student who, decades earlier, visited the fairgrounds as a child. One such fairgoer was Eva Holzapfel, of Huntington Station, Long Island, who saved enough S&H Green Stamps to visit the fair *every* weekend. Stories such as Eva's abound. And for the luckiest, when worldwide travel was still out of reach for most people, treasures such as the *Pietà*—or in Eva's case, General Electric's Carousel of Progress, complete with Disney animatronics—could be only a subway ride away.

Over time, the draw of world's fairs began to fade. Air travel became affordable, and the internet opened up virtual travel from the comfort of our homes. Yet, nothing quite compares to the memory of the excitement, the color, and the magical fantasy of a visit to the fair. It was a dreamworld in real life.

Acknowledgments

I want to acknowledge the assistance of a number of people, beginning with the archivists: Kate Feighery at the Archdiocese of New York Archives; Tal Nadan at the New York Public Library; and a special thank-you to Joseph Cohen at the Diocese of Brooklyn Archives. I am also grateful to Ed Wilkinson, editor emeritus of Brooklyn's diocesan newspaper, the *Tablet*, and to Bill Cotter, *the* authority on New York's world's fairs, for their valuable feedback. Thank you also to George Hoehmann, who graciously allowed me to read and quote his unpublished master's thesis, "'My Eminent Friend of New York': Francis Cardinal Spellman and the Second Vatican Council." I'm grateful to Nicole LaRossa, director of *Fordham News*, and John Murray Jr., both of whom assisted in supplying rare photos. I want to thank Mahinder Kingra, my editor at Cornell University Press, for his thoughtful guidance; Michael J. McGandy and Abigail Harris for their early editing; Father James Flint, OSB, for his careful review; and lastly, Rita Rosenkranz, my agent, whose knowledge and experience saw this project through from inception to completion.

Abbreviations

AANY	Archives of the Archdiocese of New York, St. Joseph's Seminary, Dunwoodie, Yonkers, 1964–1965 World's Fair Collection, Collection Number 025.002
BRTD	Jo Mielziner Papers, *T-Mss 1993–002, Billy Rose Theatre Division
CCOH	Charles Poletti interview, Columbia Center for Oral History
NYPL-CSC	Robert Moses Papers, Series 11, Committee on Slum Clearance, 1955–1959, Manuscripts and Archives Division, New York Public Library
NYPL-NYWF	New York World's Fair 1964–1965 Corporation records, Manuscripts and Archives Division, New York Public Library
NYPLPA	New York Public Library for the Performing Arts
RCDBA	Roman Catholic Diocese of Brooklyn, Diocesan Archives, Office of the Bishop

Our Lady of the World's Fair

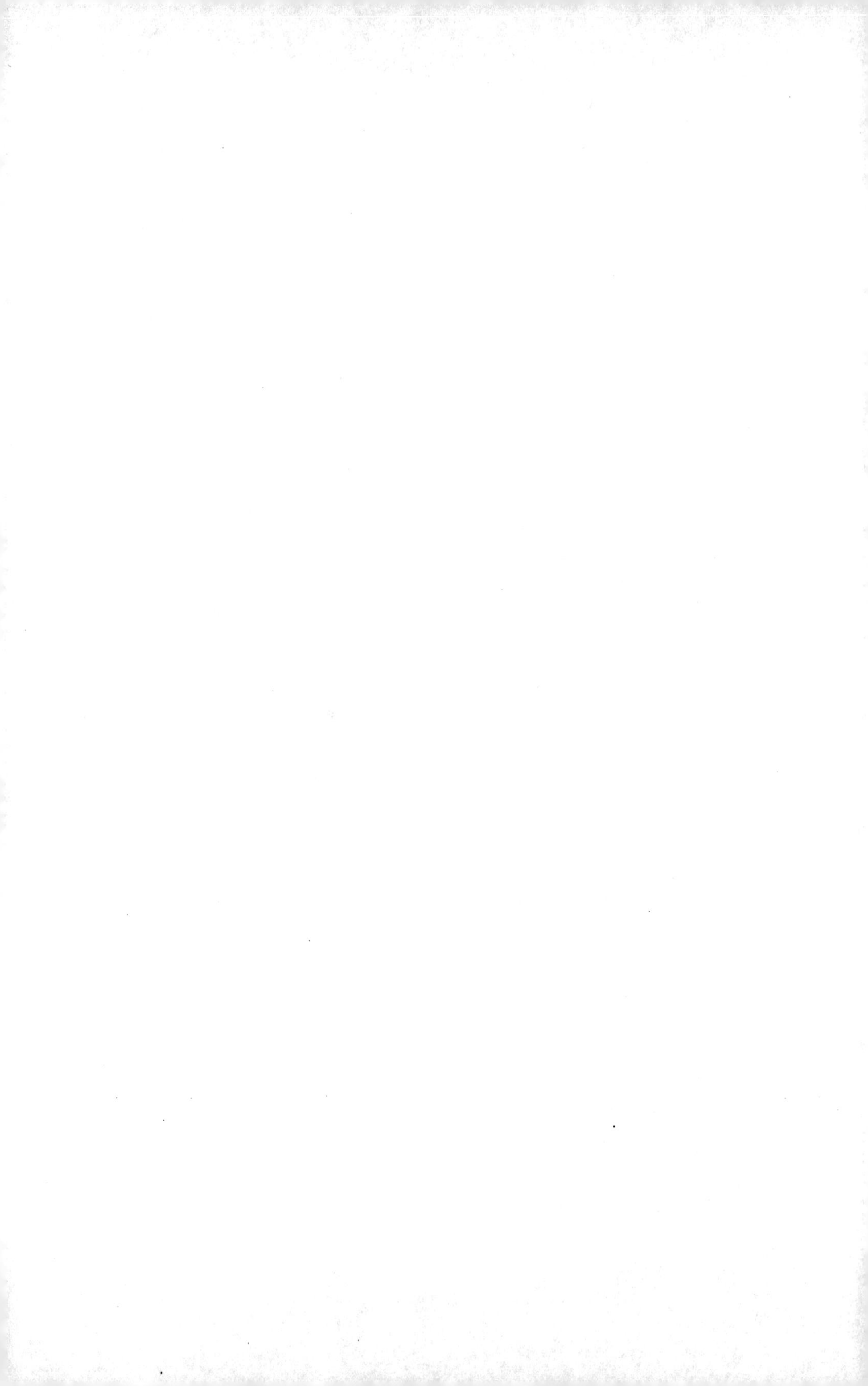

INTRODUCTION

Between April and October 1964, and again in 1965, a grand total of fifty-two million people from around the world gathered in Queens for the World's Fair. At the time, it was the most highly attended world's fair ever. Ostensibly held to honor the three hundredth anniversary of the naming of New York when King Charles II sent an English fleet to claim it from the Dutch, in truth the fair was an excuse to present a showcase of mid-twentieth-century American culture and technology. As Lawrence Samuel wrote in his retrospective, the event remains a touchstone for many American baby boomers, who visited the optimistic fair as children before the full impact of the countercultural social changes, the turbulence of the civil rights movement, and the Vietnam War was felt.[1]

The man behind the fair was Robert Moses, who more than any other figure in the twentieth century shaped the physical environment of New York. A man with unimaginable energy, he at one time held simultaneously twelve positions, all appointed. From these posts, Moses expanded the state and city park systems, built highways, bridges, playgrounds, housing, tunnels, beaches, zoos, and civic centers, and, along the way, two world's fairs. Though he was vilified by the end of his lifetime, Moses's contributions are now deemed by some as needed.[2] He was renowned not just for his ambition and seemingly unlimited energy but also for his pugnaciousness and the less than diplomatic manner in which he treated those who crossed or merely disagreed with him. There was at least one person, however, to whom Moses extended every courtesy, and that was Cardinal Francis Spellman. Moses was "Bob" to Spellman, but Spellman was always "Your Eminence" to Moses.[3]

Born to a secular German Jewish family, Moses graduated from Yale and, standing over six feet tall, presented an imposing figure. Spellman, of Massachusetts Irish ancestry, was stocky in stature, studied in Rome to become a priest, and emerged a Prince of the Church.

Spellman was installed the sixth archbishop of New York in 1939, the same year Moses opened the first New York World's Fair. Like Moses, Cardinal Spellman wielded enormous political and cultural influence in New York City—and he was as savvy about the business of the church as he was about its mission. Spellman had quickly learned of the archdiocese's alarming financial state, with mortgages totaling close to $28 million. After a thorough analysis, he undertook a sweeping reorganization of the archdiocese by centralizing, then refinancing the debt at much more favorable terms. Initially the banks balked, but when Spellman suggested moving his banking to Boston, they quietly fell in line. Pastors, too, were reluctant to surrender their fiscal control, but the savings they saw outweighed any resistance. In essence, the archdiocese became the central bank for all church enterprises.

In 1947, Spellman made his expectations clear when he introduced his $25 million diocesan building plan to the presidents of the largest construction firms and the heads of the building and trades unions. He asked for two things—"from the construction companies, a good job for the price; and from labor, a good day's work for the pay."[4] He got both, and the archdiocesan building boom was on. To understand the magnitude of this new construction program, second only to the city's, consider the five-year period from 1955 to 1959, when new construction was begun on fifteen churches, ninety-four schools, twenty-two rectories, sixty convents, thirty institutions, and hundreds of additions, alterations, and repairs, totaling over $168 million, or more than $1.6 billion in today's dollars, adjusted for inflation. Since his tenure had begun in 1939, Spellman was also responsible for building a number of Catholic hospitals, as well as expanding the high school system.

Spellman asserted his control from the office of the chancery, housed in one of six Stanford White–designed neo-Renaissance Florentine mansions, one block over from St. Patrick's Cathedral, and modeled ironically after the Roman Palazzo della Cancelleria, the fifteenth-century papal chancery palace. From this command post, nicknamed "the Powerhouse," Spellman administered what was the wealthiest Catholic diocese in the world. Unlike Moses, however, Spellman wielded influence that extended well beyond New York—it was global. In addition to his role as military vicar and a sometimes confidant of President Franklin Roosevelt, he was responsible for the church's largest foreign aid program, Catholic Relief Services, with a 1964 budget of $176 million, well over $1.5 billion today. For his administrative and financial acuity, he acquired the nickname "Cardinal Moneybags" by some in Rome, and the world was better for it.[5]

For either man to accomplish his goals, working collaboratively was a necessity. A natural alliance formed between Moses and Spellman, as contractors and trades, just like the American church, were dominated by the Irish Catholics.[6] How did the Irish ascend to such prominence? The Irish had advantages that other Catholic immigrants lacked—they spoke English, they were accustomed to financially supporting the church, and they had already experienced anti-Catholic prejudice and knew how to use their political power to fight it. For all these reasons, wrote Lawrence McCaffrey, "the Irish were the only European ethnics who could have led the American Catholic community into an accommodation with the dominant, Anglo-American Protestant ethos."[7]

Politically, Spellman had always stood firm against communist aggression, while the Vatican, led after 1958 by Pope John XXIII, Angelo Giuseppe Roncalli, had begun to move away from the hard anticommunism of Pope Pius XII. Born in 1881 to sharecroppers outside Bergamo, northeast of Milan, Roncalli was the third of thirteen children. At the age of twelve he entered the Bergamo seminary, then continued his studies in Rome. After being called to service in the Italian army, he returned to the seminary to complete a doctorate in theology and was ordained in 1904. After serving as a chaplain in World War I, he remained in Rome until he was sent to Bulgaria, then Greece as an apostolic delegate—a Vatican representative with no diplomatic status—to assess conditions, and finally to Istanbul to locate prisoners of war. As if to make up for his years assigned to obscure posts, he was appointed nuncio, the Vatican's equivalent of an ambassador, to Paris to assist in the church's efforts in France as World War II drew to a close. In 1953 he became cardinal-patriarch of Venice, where he expected to spend his last years shepherding his flock. Five years later, Pope Pius XII, after a long illness, died on October 9, 1958, and nineteen days later the College of Cardinals elected Roncalli as the next pope. Already seventy-six years old and in poor health, Roncalli was expected to be a transitional pope, and sharing that assumption, he took the name John XXIII. He selected the name of John in honor of one of the twelve apostles—but also because it was the name of a long line of popes whose pontificates had been short. Dropping his predecessor's hard line against communism, John was regarded as a reconciler. With this new openness, the Vatican agreed to participate in the 1958 Brussels Universal and International Exhibition. The first world's fair since 1939, Expo 58, as it was called, was an opportunity for Europe to showcase its technological, industrial, and cultural progress after the devastation of World War II. The theme of the fair was simply "Man," but more broadly translated, "a world for a better life for mankind," and each participating nation presented its own interpretation of mankind's journey toward happiness and fulfillment.[8]

Coming at the height of the Cold War, the Brussels World's Fair proved to be one of "the most important propaganda event[s] to be staged by European allies in the Eisenhower years."[9] Fresh off the USSR's successful foray into outer space, the Soviet exhibition was expected to make the "biggest splash," by spending $50 million, or four times more than Congress allocated for the US exhibition.[10] The Soviet pavilion showcased the USSR's advances in space exploration with a model of the first earth-orbiting satellite, *Sputnik I*, along with a display of heavy machinery and electronic technology. Reporting on the fair, American artist Howard Fussiner described it as "heavy handed . . . but not ineffective."[11] According to one critic, the Soviets "announced that they would fire point-blank in Brussels with some of their most powerful cultural batteries."[12] Though the Soviet artwork displayed was decades behind contemporary trends, Fussiner could not help but notice that "such paintings require no explanation from art critics—the meaning is at the grasp of all and it was interesting to observe the pleasure of the average visitor." By contrast, he had noticed "the embarrassed snickers or unabashed guffaws which the American abstract expressionist paintings evoked from these same peasants and workers."[13]

Since this was not a trade fair, the United States pavilion lacked any commercial or industrial displays but rather focused on the broad sweep of American life at work and play. This was a mistake. One of the main attractions was the *Circarama*, a twelve-minute Walt Disney film that was more spectacular for its delivery—it was shown on a circular panoramic screen—than for its content. At a different time and place perhaps, the "soft sell," as *Time* magazine described it, would have been entirely appropriate, but compared to the Soviets' top-gun display, the pavilion was criticized as "bewildering" and proved to be an embarrassment to the Eisenhower administration.[14]

By contrast, as Fussiner wrote in his review of the fair, "the pavilion of the Holy See had one of the very best art exhibitions of the Fair—a selection of Christian Art from its beginnings to the present, with many exquisite treasures of early medieval times."[15] *Time* magazine wrote that "Vatican City has produced one of the best shows of all."[16] While some pavilions promoted culture as their weapon in the battle against communism, the Vatican put forward the life of Jesus Christ and the message of the church. The official name of the pavilion was "Civitas Dei"—the City of God. In a report issued by Providence bishop Russell J. McVinney, assigned to sway the American bishops on participation of US Catholics in the pavilion, he emphasized that the goal was to counter the influence of the secularist world, namely the Soviet Union and its satellites, by presenting the Catholic way of life in a dignified and effective manner, since "we are now engaged in a world war for the minds of men."[17]

A team of architects led by M. Paul Rome designed the pavilion as an allegorical city surrounded by a fifty-foot-high wall with large entryways. Within the walls stood a church, described by reporters as built in the "Swedish style," which was to say, of modern design. The church interior, which seated twenty-five hundred, was outfitted by Germany and France under the direction of the architect of the Cologne Cathedral. France also furnished the small chapel located behind the church, where the Blessed Sacrament (the consecrated host) was reserved for distribution at the Mass. Five additional chapels represented the six other sacraments: baptism, confirmation, ordination, marriage, reconciliation (confession), and anointing of the sick. Luxembourg was responsible for the sacristy, the church room in which the sacred vessels and vestments are stored.[18]

The theme of the entry hall was drawn from the Old Testament, with a focus on the creation of man and the prophecies. The second hall represented the New Testament—the Christian era, featuring the life of Christ and sections devoted to the Virgin Mary and the Holy Land. The highlight of the pavilion was a reproduction of catacombs and an exhibit of a hundred sacred images, titled "Imago Christi," ranging from the fourth century AD to the present.[19]

The largest work of art in the exhibit was created by the Bratislava-born British artist Arthur Fleischmann. During his lifetime, Fleischmann received commissions from around the world, including a number of religious works. Influenced by his Catholic faith, he would eventually sculpt images of four popes.[20] The Brussels' commissioner general of the Vatican pavilion selected Fleischmann for his twenty-foot aluminum sculpture *Risen Christ*, which depicted the Resurrection. Instead of a traditional rendering of Christ, it was a more stylized, angular work, criticized at the time for being too "Germanic," yet it has proven aesthetically enduring. Working in an acrylic medium, Fleischmann also completed a two-part frieze titled *Resurrection of the People*, to illustrate man on the last day.[21]

Upper-floor displays illustrated the papacy, the Vatican, the work of the missions, Catholic charitable works, and the church's presence in contemporary society through education, social action, the arts, science, and the Vatican-sponsored press, radio, and television agencies. Traditional cast bells rang from a fifty-foot-high carillon tower, along with miniature bells operated electronically to ring out the hours and play music. As an amenity to visitors, the pavilion included a garden, which led to a glass-enclosed restaurant that seated over one thousand, with a relaxing patio terrace.[22] The pavilion represented the Church Universal, but it was dependent on American Catholics to foot a significant portion of the cost. To participate in the upcoming New York fair, the Vatican would once again have to depend on substantial donations from Americans to fund its participation.

In 1961 Pope John XXIII issued the encyclical *Mater et Magistra* (Mother and teacher), a compendium of Catholic social and economic teaching on the heels of tensions between the Italian Right and Left, and moved the church away from political involvement. Later that same year, John issued *Pacem in Terris* (Peace on earth), calling on an end to the arms race. The pope realized that his peace overtures could be misunderstood and that he risked being used by the Kremlin, but he continued to pursue a neutral path.[23] This had an impact on how the church would respond to the fight against communism in Southeast Asia.

With the American involvement in Vietnam gaining momentum, there was a growing dissonance between the pre– and the post–World War II generations, regarding not only the war but also long-held traditions and values. The 1960s witnessed the convergence of several combustible movements—the long overdue push for civil rights, the burgeoning women's movement, a shift in sexual mores, just to name a few—all leading to a cultural and political explosion. The social capital that Robert Moses and Francis Spellman had banked on for so long began to shift. The world in which both men operated was changing and changing fast. But before the crises that would soon erupt across the country, there was to be one spectacular collaboration between New York's two most powerful men. Moses leaned on Spellman to join forces with him, first in enlisting the Vatican to participate in New York's World's Fair, and then ensuring the fair's blockbuster attraction by transporting Michelangelo's *Pietà* across the Atlantic Ocean to Flushing, a neighborhood in the Queens borough of New York City where the fair would be held.

Initially, the press welcomed Moses's leadership in bringing another fair to the city. In an editorial, New York's *Herald Tribune* extolled "The Bob Moses World's Fair," adding "We can't imagine anybody who could do the job better," and labeled the coming fair as Moses's "crowning achievement in a lifetime of public service."[24] But by the time the fair started, a shift in leadership at the *New York Times* signaled that enthusiasm for Moses had began to fade. Newspapers had helped sell the 1939 fair, but twenty-five years later, their support was tempered.

Other actors were to weave their way in and out of this story. On the art scene there was John Canaday, the leading art critic for the *Times*, who not long after he arrived at the paper dismissed abstract expressionists, artists such as Jackson Pollock, whom he did not name by name but described as living in a "protracted adolescence." These artists, along with their followers, in Canaday's opinion, held an exceptional tolerance for "incompetence and deception."[25] Cultural battle lines were drawn, and with each new review—and there were many over Canaday's seventeen-year tenure—one never knew on which side he would come down. Another satellite player was the powerful monsignor from Cicero, Illinois, Paul Marcinkus, nicknamed "Il Gorilla" for his imposing muscle, which he sometimes had to exert as a papal bodyguard. The term "bodyguard," however, was inadequate to describe the roles that he held and the influence he wielded. He thought he could manage Moses, but he was wrong.

The story that unfolds in the following pages will never, nor could ever, be repeated again. It was a slice of New York history playing out from the Flushing fairgrounds to as far away as Rome. This story centers on only one pavilion, one out of one hundred and forty. It was not the largest pavilion, nor the most dazzling, but it spoke most clearly to the human spirit. It hosted the blockbuster that Moses needed to ensure the fair was a success and provide funds to complete his lifelong objective of turning a Queens dumping ground into a jewel of a park.

1

The Vision

Robert Moses and the Valley of Ashes

The documented history of the place known as Flushing Meadows stretches back to 1645, when the town of Flushing was founded on a tidal basin now known as Flushing Bay. Those were the days when townspeople could still fish, dig for clams, and hunt for ducks along the bay shore. Around 1730 Robert Prince built the first commercial nursery in town, which by the time of the Revolution was large enough to catch the notice of George Washington.[1] The surrounding marsh and farmland attracted little attention until the end of the nineteenth century, when one Fishhooks McCarthy and his Brooklyn Ash Removal Company turned it into a dumping ground for the "tin cans, cast-off baby carriages and umbrellas of Brooklyn."[2] It was a place for burning the borough's garbage, and not much had changed when F. Scott Fitzgerald described the wasteland in his 1925 novel, *The Great Gatsby*:

> About halfway between West Egg and New York the motor road hastily joins the railroad and runs beside it for a quarter of a mile, so as to shrink

> away from a certain desolate area of land. This is a valley of ashes—a fantastic farm where ashes grow like wheat into ridges and hills and grotesque gardens; where ashes take the forms of houses and chimneys and rising smoke and, finally, with a transcendent effort, of men who move dimly and already crumbling through the powdery air.

Flushing Meadows may have remained a dumping ground, known as the Corona Dump, for years to come had Robert Moses not been appointed New York parks commissioner in 1934. Born in 1888 to assimilated German Jewish parents in New Haven, Connecticut, Moses was reared in a home of material comfort and educational rigor. After graduating from Yale and furthering his study at Oxford, Moses earned his PhD in political science at Columbia University. By dint of hard work and fortuitous connections, Moses began his ascent to the highest city and state government circles. Beginning in public administration, he had a penchant for park development, and as early as 1923 Moses had conceived of a system of state parks that would stretch across Long Island to New York City, linked together by expansive parkways.[3] His success on Long Island and throughout the state sparked a state parks movement that became a model for the rest of the nation.

Public parks in America were originally conceived as restorative "pleasure parks," with early examples designed by landscape architects and gardeners such as Frederick Law Olmsted (1822–1903) and Jens Jensen (1860–1951) to offer fresh air and open space for family excursions and refreshment.[4] By the twentieth century, urban reformers saw the potential of open park space as a venue to redirect free time to more purposeful ends. Growing out of the late nineteenth-century playground movement, "reform parks" provided organized activities and recreation to protect children from the perceived detrimental effects of idleness, and eventually this vision was expanded to include adults as well.[5]

New York's golden age of reform parks came to an abrupt end with the appointment of Robert Moses as parks commissioner in 1934. Moses signaled the ideological shift from reform to recreation in his first annual report: "We make no absurd claims as to the superior importance and value of the particular service we are called on to render."[6] Recreation was now recognized as an end unto itself, and more than programming, park administrators began to emphasize park expansion and construction of recreational facilities, such as baseball fields and tennis courts.[7]

In his book on the history of the public park, *The Politics of Park Design*, Galen Cranz writes that when park departments replaced an ideology of reform with one of leisure, they "put themselves on a par with commercial producers of entertainment commodities."[8] In retrospect, converting a city park to a world's fairground—one of Moses's final public works projects—was pragmatic.

By the 1940s, after decades of piecemeal city planning, cities began to recognize the need for greater coordination between various agencies. Since the days of Olmsted, park professionals had pushed for a more comprehensive approach to urban planning, calling for the adoption of a "master plan."[9] Moses recognized this need and began consolidating New York's five borough park bureaucracies. Not limited to parks, Moses left his mark on a wide range of building projects, including 13 bridges, 2 dams, 568 playgrounds, and numerous prominent public buildings, including the Lincoln Center for the Performing Arts, Shea Stadium, and the United Nations Headquarters building. Roberta Brandes Gratz, in her book *The Battle for Gotham*, writes that Moses accomplished this by molding the nascent institution, the public authority, into an independent agency that functioned outside the checks and balances of the government, the press, and the public eye.[10]

In Moses's obituary in the *New York Times*, Paul Goldberger wrote that Moses "held 12 public offices simultaneously, ranging from head of the Triborough Bridge and Tunnel Authority to head of the State Power Commission to one all-embracing post created especially for him, City Construction Coordinator." Goldberger observed that Moses seemed "to be less in debt to governors, mayors and even Presidents than they appeared to be to him."[11] Part of this influence may be attributed to Moses's long stay in power, which began in 1924 and ended more than four decades later in 1968, outlasting five mayors, six governors, and seven presidents.

Yet despite the scope and magnitude of his accomplishments, even during the peak of his power critics began to chip away at Moses's reputation. Writer and activist Jane Jacobs's 1961 book, *The Death and Life of Great American Cities*, resonated with New Yorkers who saw their neighborhoods wiped away. Thirteen years later, Robert Caro's Pulitzer

Prize-winning biography, *The Power Broker*, was the coup de grâce. Caro begins by describing Moses as "perhaps the single most influential seminal thinker" in the field of urban planning and quotes Lewis Mumford, Moses's most bitter critic, who said, "In the twentieth century, the influence of Robert Moses on the cities of America was greater than that of any other person."[12] Yet in Caro's exhaustive 1,246-page biography, Moses was held responsible for the demise of the city of New York. Moses was discredited for his "planning by fiat" style; in particular, Caro criticized him, as did many others, for his approach to the then-delicate turn of phrase "slum clearance." To that Moses replied, "I raise my stein to the builder who can remove ghettos without removing people as I hail the chef who can make omelets without breaking eggs."[13]

While many of the urban renewal projects Moses led were needed—that much was agreed—it was the way in which they were executed that alarmed community leaders. The destruction of entire communities fell disproportionately on Hispanics and African Americans. Driven by a "lava flow" of federal funds, redevelopment and expressway projects wiped out hundreds of businesses and displaced thousands of mid- and low-income residents, aggravating an already tight housing market.[14] The social and financial fallout of this mass displacement continues to reverberate today. While Moses's record is still debated, in the light of delays and difficulties in recent huge municipal and state projects (such as New York City's 9/11 Monument), some urban planners are again reconsidering his legacy.[15]

With an extraordinary talent for grasping the bigger picture before anyone else, Moses conceptualized the Flushing eyesore as a bigger and better Central Park. The meadows of Flushing totaled 1,346 acres, one and a half times the size of the park in Manhattan. His plan was to purchase the land on either side of the Grand Central Parkway that ran through the dumping ground, but that was beyond the city's resources. Fifteen years after he first conceived of transforming the dump, his park was still not realized. When the idea to hold a fair in 1939 was proposed, Moses moved quickly to champion the venture. However, he let the organizers know that Flushing Meadows was the only place in New York where they could get any cooperation from the Parks Department while he was in charge.[16] As parks commissioner, he, in effect, would become the fair's landlord.[17]

Responsible for preparing the site for the 1939 fair and improving the roadways, Moses leveled the ash dump, filled in much of the meadow, created two lakes north of the filled land, built new traffic arteries, and installed permanent utilities for the park and temporary utilities for the fair. Two giant sewage treatment centers had to be built just to clean out the pollution in Flushing Bay. Moses had the grounds landscaped and two permanent buildings constructed: the New York City Building, which served as a skating rink after the fair closed, and the State Auditorium, intended for later use as a concert venue. All told, the total cost for improvements amounted to a staggering $59 million.[18]

Setting a futuristic theme, the motto of the 1939–1940 New York World's Fair was "The World of Tomorrow." The theme was fitting, as the fair introduced many technological advances that would come to define the quintessential American lifestyle for the next fifty years, such as the newly invented television. The World's Fair Corporation reinforced this modernist vision by constructing pavilions and exhibitions that visually conformed to its art deco design program.[19] After years of scarcity following the economic collapse of the Great Depression, the fair offered a glimpse of the economic and technological advances that would usher in the comfort and prosperity of the following decades (but only after another world war, as it turned out). It was against this background that expectations for the 1939–1940 fair were established.

In recent years, scholars such as Jesse T. Todd have approached world's fairs as if they were "petri dishes" where microcosms of developing national cultures and identities could be studied.[20] Missing from many retrospectives of the 1939 fair is any mention of the faith-based pavilion, the Temple of Religion. This was not the first world's fair in which religion was on exhibit. In his study of religion-sponsored exhibitions at America's pre–World War II fairs, Todd notes that Chicago's Columbian Exposition of 1893, the Philadelphia Sesquicentennial of 1926, and Chicago's 1933–1934 Century of Progress all included exhibits highlighting the contributions of "churches and religious groups to nation-building, and celebrated religious freedom and tolerance as markers of modern progress."[21] What made the 1939 pavilion unique was its ecumenical nature.

Heavily subsidized by John D. Rockefeller Jr., who oversaw the distribution of his father's fortune, the pavilion was an unprecedented collaborative effort among Catholics, Protestants, and Jews. The dedication of the Temple of Religion was the first official ceremony of the opening of the fair, with both Governor Herbert Lehman and Mayor Fiorello La Guardia addressing the twelve hundred invited guests. Rockefeller, as the keynote speaker, remarked that he considered this pavilion to be "one of the central buildings of the fair."[22] According to one later account, the pavilion received as many as 4.5 million visitors during the fair's run.[23] The popularity of the tri-faith exhibition would anticipate a strong religious presence at the 1964 fair, again with generous support from the Rockefellers.

Marc Miller wrote in his analysis of Moses's role in the 1939 fair that in contrast to his future high-stakes involvement in the 1964 fair, Moses was "nearly indifferent to the fair itself, interested only in the event's potential to generate permanent civic improvements."[24] Moses viewed the fair as a means to an end; it was a business- and government-supported enterprise that would transform the fairgrounds into a permanent park: the "Versailles of America." Although in retrospect grandiose, Moses's vision contained a well-conceived infrastructure plan. At the geographic center of the New York City metropolis, Flushing Meadows had the potential to become a logistical hub for a network of roads and bridges. All future improvements would be financed from the fair's profits—but there were no profits. Owing to mismanagement by the director, Grover Whalen, the fair closed with a loss of $200 million.[25] True to form, Moses was neither deterred nor discouraged.

In 1946, with the end of World War II and when the newly established United Nations needed a headquarters, Moses persuaded the organization's first secretary-general, Trygve Lie, to use one of the two permanent buildings from the fair, the New York City Building, as its temporary home through 1950. That decision brought in an additional $2 million for remodeling the building and landscaping the park. Moses pulled in other public works projects, always with an eye to completing Flushing Meadows Park. Years later, when a second world's fair was proposed, ostensibly to commemorate the three hundredth anniversary of the British acquisition of the village of Nieuw Amsterdam from the Dutch and its renaming to New York, Moses saw the realization of his dream finally within his grasp.

Moses's Vision for the 1964 World's Fair

Capitalizing on the existing infrastructure at Flushing Meadows, in 1959 New York successfully lobbied President Eisenhower to approve its "cheaper-than-cheap" site for the next world's fair.[26] A new Flushing stadium was already a twinkle in Moses's eye when he had earlier offered a new home for the Brooklyn Dodgers to team owner Walter O'Malley. O'Malley declined, hoping instead for a new domed stadium in Brooklyn, but Moses denied him this option, leading to the team's relocation to Los Angeles.[27]

Moses angled his way into the role of president of the World's Fair Corporation, which entailed resigning as parks commissioner to avoid a potential conflict of interest. He took charge of every aspect of the fair—no detail was too small.[28] His brash and forceful approach, which worked so well in New York, offended the members of the Paris-based Bureau of International Expositions (BIE), whose approval Moses needed for official recognition of the fair. Most European countries, thirty to be exact, were members of the BIE, formed in 1928 to supervise international exhibitions. Since the first Great Exhibition held in London's Crystal Palace in 1851, the popularity of world's fairs had grown to such a degree that by the early twentieth century competing fairs were being held concurrently. To regulate the quality and ever-growing numbers of exhibitions, the BIE, an intergovernmental organization, established a two-tier classification for world's fairs: Category One, the "universal" higher tier, and Category Two, the "international" fair, both referred to as "Expos." A universal exposition was a world's fair in the "grand tradition," and one that was "characterized by a far-reaching theme that has to do with several branches of human endeavor," writes Martin J. Manning in his analysis of US participation in world's fairs.[29] For a fair to qualify as "universal," the BIE required a substantial number of international exhibitors, from the national government level, state and provincial level, down to cities, corporations, and public-interest organizations. It must draw more than five million attendees, run six months, and be centrally located with "contiguous spaces for exhibits and public gatherings."[30] By all accounts, Moses's fair would qualify for universal status. However, only one universal exposition was allowed to be held within a ten-year period, and the

1962 Seattle World's Fair was already approved.[31] That was just one of the BIE's objections.

From his experience with New York's first world's fair in 1939, Moses understood that the fair would have to run for two years to realize a profit; but the BIE only permitted one six-month run. Additionally, Moses wanted to charge rent for pavilion lots, another violation of bureau rules, which mandated rent-free lots for national pavilions. In the past, the BIE had made exceptions, as in the case of the 1939 New York World's Fair, which was formally contracted for one year, on the condition that the United States would become a member nation. US membership in the BIE would have required an act of Congress, and there was no appetite in Congress to give Parisian bureaucrats the power to direct Americans; furthermore, membership cost money. In any case, back in 1939, with World War II looming, Congress had more urgent matters with which to deal.[32]

With the advent of the Cold War, President Eisenhower established the United States Information Agency (USIA) in 1952 to coordinate the US government's information outreach overseas—essentially a public relations branch to counter communist propaganda. The USIA embarked on a wide range of efforts and promotions, such as the Voice of America radio broadcasts, overseas libraries, publications, and exhibitions. The Brussels Universal Exposition was the first Category One sanctioned fair held after World War II, and the first overseas exposition in which the USIA was responsible for US participation. In 1961, Congress expanded and consolidated international exchanges, including participation in expositions, by passing Public Law 87–256, the Mutual Educational and Cultural Exchange Act (Fulbright-Hays).[33] As Manning further writes, "What the legislation did not promise was continual Congressional funding for these fairs; this had been one of the United States' inherent problems at all fairs. Unlike other BIE members, the U.S. is the only signatory country [the US did join the BIE, with significant reservations, in 1968] that does not subsidize its domestic fairs . . . and depends heavily upon private sponsorship to supplement Congressional funding for those fairs overseas."[34]

Already burned once by a US promise to join, the BIE refused to make another exception, so it withheld endorsement of the New York fair.[35] As president of the New York World's Fair Corporation, Moses had as his first priority obtaining commitments of participation from governments around the globe. Spurned by the BIE, Moses decided to circumvent the organization altogether and appeal to European governments directly. This prompted the BIE to take the unusual step of requesting its members not to participate.[36] It did not help that—as picked up by the *New York Times*—Moses described the BIE, while he was leaving their Paris office, as "three people living obscurely in a dumpy apartment in Paris."[37]

Anticipating the absence of major foreign exhibits and the financial support they would provide, Moses increased efforts to recruit smaller nations, as well as corporations. In the summer of 1960, with the hoped-for opening of the fair still more than four years away, Moses led a committee of five men on a European roadshow. The committeemen were Peter Grimm, a former chief of the US Operations Mission, a Cold War security office in Rome; John F. Simmons, vice president in charge of foreign relations and exhibits for the fair; William A. Berns, the fair's director of communications; and Thomas J. Deegan Jr., chairman of the executive committee of the World's Fair Corporation and vice president for international affairs and exhibits. A major asset to the team, Deegan was a local attorney who would later become an adviser to Lyndon B. Johnson during Johnson's 1960 presidential campaign (which Johnson would lose to John F. Kennedy). The fifth member was Charles Poletti, who had briefly served as governor of New York. Poletti was a known figure in Italy who had led efforts to restore the civilian government in Sicily. He was honored by the Catholic Church, decorated as a Knight Grand Cross of the Holy Sepulchre by Pope Pius XII and a Knight of Malta—this despite not being Catholic.[38]

Through these men's efforts, the fair would eventually host 140 pavilions representing eighty foreign countries, twenty-four US states, and over forty-five corporations, including a swath of Fortune 500 companies, such as Johnson's Wax, Eastman Kodak, IBM, and of course, the "Big Three" automakers General Motors, Ford, and Chrysler. While some countries, such as Austria, Sweden, and Spain, found a way around the restriction by participating "unofficially" through their country's chamber of commerce, some BIE-member countries, such as Lebanon, would simply set up a government-sponsored pavilion "without recourse to any dodge at all."[39] Other nations previously locked out of or unable to participate in earlier fairs were only too happy to join. These were primarily nations from South America, the Middle East, and Africa, including Argentina, Brazil, Ethiopia, Saudi Arabia, Syria, and even Sudan. Other non-BIE-member states agreeing to exhibit were India, Indonesia, Ireland, Mexico, and Japan. The Soviet Union had originally committed to participating but, owing to financial considerations, coupled with State Department interference, backed out. Faced with the absence of the Soviet Union, Britain, and even Italy, Moses set his sights on the Holy See.

The six-man team had already visited Yugoslavia, Austria, Greece, and Turkey by the time they arrived in Rome on September 1. There the team conferred with government and Vatican officials, and a papal audience was scheduled. Moses wanted to personally meet with Pope John XXIII to invite the Holy See to participate in the fair. The pope's emphasis on *aggiornamento*, or "updating," would dovetail nicely with the Vatican's presence at the New York World's Fair. Furthermore, the fair would only be a success, in Moses's opinion, if the three most powerful organizations in the world—the US government, General Motors, and the Vatican—took part.[40] The first two were a sure thing, so Moses now had to focus his efforts on winning over the pope. To do this he would have to lean on the one person who could deliver: Francis Cardinal Spellman.

2

Francis Cardinal Spellman

Faithful Son of America and Rome

When named a cardinal in 1946, Francis Joseph Spellman, both diplomat and pastor, became the most formidable Catholic churchman in America as the leader of the nation's forty-five million Roman Catholics.[1] Spellman's remarkable life began in 1889 in Whitman, Massachusetts. Of Irish heritage, his mother was described as loving and indulgent. His father, a successful grocery owner, took his son's early achievements in stride. "Son," he used to say, "always associate with people smarter than yourself, and you will have no difficulty finding them."[2] A school of distinction then as it is today, the North American College, the American seminary in Rome, admitted Spellman after completion of his degree at Fordham University in 1911. Five years later, Spellman was ordained and returned home with a doctorate in hand to begin his ministry as a priest in the Archdiocese of Boston. As if to diminish him, his first assignment was a chaplaincy at a home for aged women; then in due time he received his parish assignment. With America's entry into World War I, Father Spellman was accepted as an army chaplain, but just prior to his departure, his superior, Cardinal O'Connell, with whom he never found favor, abruptly appointed him to the staff of the archdiocesan newspaper, the *Pilot*, with instructions to increase subscriptions. Despite what was surely another blow, Spellman took to his assignment with enthusiasm and spent the next four and a half years visiting every parish in the archdiocese to promote the paper. He not only increased subscriptions but used the position to sharpen his skills by studying journalism and writing an occasional editorial.[3] In what would become a characteristic of his life of service to the church, Spellman seemed determined not merely to meet expectations but to surpass them.

After nine years in Boston, Spellman was appointed an official of the Congregation for the Extraordinary Affairs of the Church under the Vatican secretariat of state. Given the honorific title of monsignor, he was the first American to hold a position in the Roman Curia, the administrative body of the Holy See. In this curial role, Monsignor Spellman made a number of lasting friendships, including with Eugenio Pacelli, the cardinal secretary of state. Pacelli had a dim view of Americans but respected Spellman as an equal and shared his distrust of communism in general and the Soviet Union in particular. So trusted was Spellman by the Vatican that he was enlisted to smuggle Pope Pius XI's antifascist encyclical, *Non Abbiamo Bisogno* (We do not need), out of Rome for publication in Paris in 1931. Pius XI named Spellman auxiliary bishop in Boston in 1932, which positioned him to move up the hierarchy upon the death of New York's Cardinal Hayes. Pius XI, however, selected the archbishop of Cincinnati to fill the opening—but died before he signed the designation papers. In what must have appeared as the mighty hand of God, Cardinal Pacelli, Spellman's longtime friend, was elected as the next pope, Pius XII, and two months later he appointed Spellman as the next archbishop of New York, then the nation's most prominent diocese, consisting of Manhattan, Staten Island, and the Bronx in New York City, plus Westchester, Putnam, Dutchess, Orange, Rockland, Sullivan, and Ulster Counties.[4]

At the outbreak of World War II, President Roosevelt selected Spellman as military vicar to the armed forces, overseeing the spiritual services to all Roman Catholics in the US armed forces. During this period, Roosevelt also appointed the archbishop to serve on the American Commission for the Protective Salvage of Artistic and Historic Monuments in War Areas, otherwise known as the Roberts Commission, the original "Monuments

Men." This commission was charged with the recovery and protection of European art treasures from Nazi art thieves. Although not always in agreement, Roosevelt and Spellman met on a number of occasions. Spellman was well received by Eleanor Roosevelt, but this friendship became strained over the issue of federal funding of parochial schools, which she opposed.

Pius elevated Spellman to the College of Cardinals in 1946, and his tenure in that high office coincided with a rise in the affluence of his parishioners. Like most immigrants, Catholics arrived in this country at or near the bottom of the economic ladder. But by the beginning of the 1960s, the average income of most Catholics approached the national median. Officially, there were about 2.5 million Catholics in all of New York City, although more than half lived in Queens and Brooklyn, which formed a separate diocese. Still, there were larger Catholic jurisdictions in other cities with heavier concentrations of Roman Catholics. Chicago's share of Catholics was 42 percent of the overall population, and Boston's 50 percent, compared to New York's 31 percent, which was close to the city's Jewish population of 27 percent.[5] Nevertheless, as cardinal of the largest city in the United States, Spellman was now the symbol of the American Catholic Church. With influence at home and abroad, Spellman worked to improve relations between the church and the Jewish community, supporting Israel's entry into the United Nations in 1949. He was a vocal opponent of antisemitism in the church, and in a city with a large Jewish population, that went a long way.[6] Henry Morton Robinson's best-known novel, *The Cardinal*, published in 1950, was loosely based on the life of Spellman. In December of that year, *Time* magazine reported that it was "the year's most popular book, fiction or nonfiction." The novel inspired the 1963 Academy Award–nominated movie by the same name, directed by Otto Preminger, with Tom Tryon playing the lead.[7]

Few people—not even Spellman—could have foreseen the changes that would soon be coming to the church. In 1959, only three months into the pontificate of John XXIII, the seventy-seven-year-old pope had announced his intention to hold a Second Ecumenical Council of the Vatican, a general or universal assembly of bishops to discuss the Catholic Church in the modern world. Between 2,000 and 2,500 bishops and many more observers and laypeople were expected to attend four sessions at St.

Peter's Basilica between 1962 and 1965. Commonly referred to as Vatican II, the council worked to renew Catholic doctrine for a modern world and restore unity among all Christians. The impact was profound, with so many traditions altered or completely dropped. The changes were sweeping, none more than the use of the vernacular, in place of Latin, for the celebration of the Mass.

As the senior American cardinal, and labeled as the obvious "conservative," Spellman was to play an important role in the Vatican II council.[8] George A. Hoehmann, however, in his study of Spellman and his participation in the council, described a more nuanced man. In particular, Spellman brought in the brilliant yet once-silenced Jesuit theologian John Courtney Murray as a *peritus*, or adviser, to champion religious liberty. Hoehmann writes:

> In summary of Cardinal Spellman at the Vatican Council two distinct points must be made. First, Spellman cannot be pigeon-holed as some reactionary conservative idealogue—he was multi-facetted. On the one hand he was both "conservative" and "progressive," staunchly "American" and at the same time "Roman." During the early days of the first session he forcefully intervened concerning religious liberty, and would be considered to have been in the mainstream. However, during the final days of the fourth session he bitterly opposed . . . the Pastoral Constitution on the Church in the Modern World[,] and [in] the treatment concerning nuclear weapons, he was in the embarrassing minority. Spellman cannot and should not be pigeon-holed.[9]

When it came to the issue that most immediately impacted the Catholic in the pew, Hoehmann adds that "Spellman explicitly asked that Latin be retained as the language of the Western Church in her liturgy."[10] While many of the Vatican II changes were "prudent and necessary accommodations to the times," they didn't strike everyone as improvements.[11] Latin was swept away, and the Mass was now to be celebrated in the vernacular. Gregorian chant was out, and contemporary songs with catchy tunes were introduced to permit greater participation in the worship service. While Spellman applauded many of the achievements of the council, he struggled with these changes and said, "In my opinion the [liturgical] changes are too many and too fast."[12] However, Spellman complied: "I don't like it," he was quoted as saying, "but I am a loyal son of the Church and will implement every one of the changes."[13]

The council addressed a number of other issues, such as the demographic shift from a church focused on Europe and the Americas to one that embraced the growing church in Africa and parts of Asia. It took up ecumenism with more openness to Protestants and Jews and paid attention to the rise in secularism across Western societies. Vatican II, however, was silent on communism, although the Soviet Union was at the time at the height of its powers. The absence of any reference to or condemnation of communism in the council's documents came as a surprise, given the church's staunch condemnations of communism under Pope Pius XII, who until his death in 1958 was unwavering in his denunciations of communism.[14]

Spellman, who was anticommunist to his core, was not so reticent, and it is through this lens he can be best understood. As military vicar for the armed forces—in itself one of the largest dioceses in the world—Spellman had a particular vantage point from which to observe, as he wrote, "the spawn and spoils of Nazi-Fascist-Communistic totalitarianism." He continued, "I saw them in the refugee camps and in the roadside gutters of the war-ruined world, and from their livid lips I learned terrible testimony to the mockery of the totalitarianism 'paradise' from which they had been liberated! They told no tales of Communist plenty, equality and justice. Theirs were stories of hunger, oppression, and death."[15]

Spellman saw a type of complacency in which many Americans considered communism as a mere nuisance, or worse, "a philanthropic movement to improve the condition of the poor and underprivileged."[16] Communism had made inroads in the United States during the Depression, as many Americans became disillusioned with capitalism and some found communist ideology appealing, even trendy—often finding themselves later to be duped. Moreover, Spellman was dumbfounded after details of the 1945 Yalta Agreement became known, revealing that Roosevelt and Churchill had made concessions to Stalin, enabling his takeover of Eastern Europe. Spellman had always enjoyed good relations with Roosevelt, but now felt the president had not only jeopardized the security of the United States but also imperiled the Catholics trapped behind the Iron Curtain.[17] Persecutions of Catholics had followed in Poland, Czechoslovakia, and Hungary, where show trials of Catholic prelates, such as

Archbishop Stepinac and Cardinal Mindszenty, both known personally by Spellman, confirmed his every fear. And as fixated as Spellman was on the Soviets, they were equally obsessed with him. After the Mindszenty trial, a pattern had now been established whereby the name of Cardinal Spellman was dragged in and used against the defendant in almost every important staged show trial.[18] Communism was now "the single most powerful institutional opponent of Roman Catholicism,"[19] so it should be no surprise that Spellman gave his full blessing to the war in Vietnam. This support cost him when much of the country's mood turned against the war. But before America became involved in the Southeast Asian conflict, Spellman had ties to Vietnam that had extended decades earlier with the support of Vietnam's Catholic population and their bishop, who was the brother of South Vietnam's controversial president Ngo Dinh Diem.

Spellman supported US involvement in the Vietnam War, which he viewed as "a war of civilization" and justified as self-defense in the face of communist aggression.[20] The Soviets, for their part, now labeled Spellman as "one of the most ominous figures in present day America."[21]

Nevertheless, as antiwar protests spread in the United States and as the movement for civil rights gained momentum, Spellman proved more enlightened and sympathetic to opposing views than did Moses. Thomas J. Shelley observed in his history of the archdiocese that many clergy of Spellman's era had limited firsthand knowledge of African Americans and their political and socioeconomic circumstances, but that did not mean they were bigots. From the beginning of his tenure, Spellman ensured that parishes and parochial schools in Harlem were not slighted and that all schools were to welcome both black and white students. He acknowledged "the sins of segregation and discrimination" and thanked the African American faithful for not succumbing to communism's exploitation of this injustice. Shelley continues:

> With that background, it is not surprising that Spellman was bewildered by the emergence of the civil rights movement in the 1960s. However, despite his conservative instincts, Spellman began to take a more sympathetic and assertive position on civil rights. On the occasion of the more famous march from Selma to Montgomery, Alabama . . . [in 1965] Spellman issued a pastoral letter in which he described "racial and civil injustice [as] a cancer attacking the very life of our national society."[22]

In the pursuit of civil rights for African Americans, he declared that "no price is too great to pay, no sacrifice too painful, no labor too demanding,"[23] but Spellman also viewed the civil rights movement as part of a broader struggle to stave off communism. He responded similarly in the postwar years to the waves of Puerto Ricans migrating to New York. With an influx of several hundred thousand Puerto Ricans, Spellman initiated a major push for their spiritual care that has been referred to as "one of the most impressive pastoral achievements of his tenure."[24]

Spellman and Moses

In many ways Francis Spellman and Robert Moses were similar men, and their rise to power ran in parallel tracks; Spellman was appointed archbishop in New York in 1939, the same year Moses saw the opening of the city's first world's fair. Both were visionaries whose critics constituted a small industry. If Moses was the city's "master builder," Spellman was the church's. A story circulated among New York's Catholics that when Spellman presented himself to Saint Peter at the gates of heaven,

> Peter [asked him] his name and a circumstance or two that might commend him. "Francis Cardinal Spellman," was the answer, "but I prefer to think of myself as a simple parish priest who had a million and a half souls in his care." St. Peter excused himself, explaining that he would have to consult his files. When he returned, it was to say that he had been unable to find an entry. "Perhaps," he suggested, "Your Eminence's good works have been listed under another category?" The applicant then said, "I published meditations and prayers, poems and a novel, all having religious themes." St. Peter again receded from view, to return in confusion. "I'm afraid," he said, "that your writings, however commendable, have escaped our records. Was there not yet some other activity that should have attracted our notice?" With some acerbity, the Cardinal said, "Among other things, I built fifty churches, two hundred schools, as well as hospitals, homes for the aged and afflicted, religious houses, and other establishments too numerous to count, all in the service of the church." Making sure that the gate was still locked, St. Peter vanished a third time, but when he came back he was in a state of overflowing apology. "Come right in, Frank," he exclaimed. "We had you under Real Estate."[25]

Like Robert Moses, Francis Spellman transformed the landscape through an ambitious building project, in his case including churches, schools, and hospitals for a congregation that had doubled to almost two million from the start of his tenure. Both men needed the other's cooperation to accomplish their building goals. Moses's biographer Robert Caro writes that "sometimes [Moses] and the Church swapped pieces of land as casually as if they were playing monopoly."[26] It was not that simple.

Fordham University and the Lincoln Square Project

Moses's building projects were accelerated through the Title I program of the US Housing Act of 1949, which offered deep federal subsidies for clearance of slum areas to promote private redevelopment. Through the Mayor's Committee on Slum Clearance, Moses, as the body's chairman, fully capitalized on the tens of millions of dollars available to demolish slums and rebuild blighted buildings. The major resistance stemmed from residents of these communities who found themselves displaced. In Moses's drive to redevelop targeted neighborhoods, not only families and businesses were forced to relocate, but Spellman had to cede Catholic churches and school buildings. This occurred often enough that Spellman formed the Committee on Housing and Urban Renewal to work to preserve neighborhoods and stabilize parish life caused by dislocations. If that were not possible, the committee tried to ease the impact on parishes in the path of new construction projects.[27] Losing a parish could not have been welcomed by Spellman, as in the Lincoln Square Title 1 project, the largest and most ambitious redevelopment site and the crown jewel in Moses's Title 1 portfolio.

As initially conceived, the Lincoln Square Title 1 project was to encompass a twenty-five-acre tract north of Columbus Circle, from West 60th to 69th Street, and from Broadway to West End Avenue. The bulk of the project was to be dedicated to a new performing arts center, today known as Lincoln Center. In addition, two blocks were set aside for the establishment of Fordham University's Manhattan campus. Originally founded by Bishop John Hughes in 1839, Fordham University was established in the Bronx neighborhood of Fordham, for which it was named. Six years later, the French Jesuits purchased the school and have led it ever since, producing some of the country's leading politicians, business, and cultural illuminati.[28]

Among the university's alumni was Cardinal Spellman, whose relationship with Moses may have influenced the decision to include Fordham in the Lincoln Square Title 1 Project, as it was characteristic of the cooperation between the two men. Other educational institutions such as New York University, New York University Medical Center, Long Island University, and Pratt Institute had already benefited from previous Title I projects, and Fordham was an obvious candidate.[29] Based on archival documentation, Spellman's role in this project was peripheral. When faced with early opposition to the Fordham campus from prominent Catholics, such as the Manhattan borough president, Hulan E. Jack, and the city comptroller, Lawrence Gerosa, Moses informed Spellman by a hand-delivered letter, simply stating that support for the project was lacking and he would "drop out" and leave the project to others. Spellman got the hint, and the matter was swiftly resolved. Spellman was more concerned with the loss of yet another parish, St. Matthew's, at 216 West 68th Street, founded in 1902. Spellman requested and received from Moses property within the new project borders on which to rebuild the church. But as protest arose against the inclusion of Fordham on the grounds that it was a violation of church-state separation, Spellman withdrew his request.[30]

Eventually reduced to a sixteen-acre complex at the intersection of 63rd and Columbus Avenue, this project was, as architectural and urban historian Hilary Ballon asserts, "absorbed into the fabric of the city." But, she continues, "Moses put the interest of the city over those of the neighborhood," which would eventually be his undoing, as New Yorkers demanded a voice in the future of their communities. But with the advantage of a historical distance of fifty years, Ballon validates Moses's confidence in art centers and universities serving as engines of redevelopment.[31]

Illustrating further the rapport between the two men, Thomas J. Shelley, in his history of the Archdiocese of New York, recounts how the archdiocese acquired a building on Lower Manhattan's State Street where Elizabeth Ann Bayley Seton once lived. Seton, born in New York in 1774, converted to Catholicism, founded a religious order to serve the poor, and established the first parochial school in the United States. In 1963, she was the first person born in America to be canonized by the Catholic Church, and in honor of her beatification, Spellman built a Georgian-style chapel on the site of the former mission and was planning to demolish the original residence next door. In an uncharacteristic reversal of roles, Moses urged him not to. Today the building is a New York City Landmark, listed on the National Register of Historic Places, and serves as the rectory of the Shrine of Saint Elizabeth Ann Bayley Seton.[32]

These two powerful men had different approaches and temperaments but shared common values and found common cause in not only building a successful world's fair, but one that could advance their own goals, whether spiritual or temporal. In the early stages of planning the fair, Moses brought Spellman to the table to gauge his support and to secure his cooperation. He would need every bit of both.

3

Moses and Spellman

The Big Favor

As anyone who has traveled to Rome in September knows, it can be brutally hot. Robert Moses led his executive team to Rome to meet with Pope John just outside the city at Castel Gandolfo, which is actually not a castle, as its name suggests, but a villa designed for Pope Urban VIII in 1626 by Carlo Maderno. It was, and still is, customary for Romans to retreat from the urban center to avoid the summer heat and what was considered unhealthy air. The papal villa was built on the site of the Roman emperor Domitian's villa, and except for a period at the turn of the last century, Castel Gandolfo has been the official summer seat of the pontificate to the present day. Designated as an extraterritorial property of the Holy See, the palace served as a shelter for Jewish refugees during World War II.

The American visitors spoke in English, which the pope understood, but he replied in Italian, with a translator providing an English translation, although team member and former governor Charles Poletti, who also spoke Italian, could have done the job. When Moses announced to the holy father that the fair's official theme would be "Peace through Understanding," the pope was immediately responsive. The *New York Times* reported that he asked how many attendees were expected and was impressed to hear it would be seventy million. There would also be parking around the fairgrounds for twenty thousand cars—now *that* was truly impressive.[1] John spoke more freely than previous popes and could even throw in a joke. Poletti invited the pope to the fair, adding that "we've got good restaurants, and you can get fine fettuccine, and you can get some fine risotto, and we have good Italian wine." The pope simply responded, "I don't like to travel . . . *ormai, quei tempi sono passati* [alas, those days are gone]."[2] Before the group left, Pope John said a special prayer for the members of the delegation. And with the twenty-five-minute meeting at a close, the Vatican was in.

There were still details to work out, namely financing, which the Vatican secretary of state, Domenico Cardinal Tardini, and the governor of Vatican City, Count Enrique Galeazzi, would manage.[3] In 1958, the Vatican pavilion had produced sufficient funds to cover its costs, with the help of a joint effort of fifty-three countries. This fair, however, would be much costlier and, unlike Brussels, would be on the other side of the Atlantic. It would fall to Cardinal Spellman, the leader of the Archdiocese of New York, to come up with a plan, for which he was eminently qualified.

Unlike the previous 1939 World's Fair, where the Fair Corporation carried the cost and responsibility of building the pavilions, in this fair every exhibitor would build its own pavilion. By Cardinal Spellman's own admission, he was not initially enthusiastic about hosting a Vatican pavilion at the fair. In his own account of events, he flatly did not want to assume this responsibility, and when he brought the matter before the Administrative Board of Bishops, they too were "unanimously opposed to the idea because of its cost, and while the American Bishops had contributed to the Vatican Pavilion in Brussels, they knew they could not have any help from Belgium or any other country in this project. The Vatican was disappointed," Spellman recounted, "and I was asked again to interest myself in the matter."[4]

On December 7, 1960, William E. Potter, supervisor for all construction for the fair, bumped into Cardinal Spellman at the Lotus Club. Located at Five East 66th Street, the literary club was founded in 1870 and was referred to by Mark Twain as simply the "ace of clubs."[5] Housed in a French Renaissance–style brownstone designed by Richard Howland Hunt and completed in 1900, the "Club" was, and remains, an exclusive meeting place for New York's elite. Potter had just completed his thirty-eight-year tenure in the Army Corps of Engineers as governor of the Panama Canal Zone. During World War II he had served as an engineering officer at the Supreme Headquarters of Allied Powers Europe (SHAPE) and helped plan the invasion of Normandy in 1944. His military experience established him as a person with excellent credentials for overseeing the building of an array of national and corporate pavilions. Walt Disney would later hire Potter to oversee the construction of Disney World in Florida, but for now he was one of "Moses's Men."

Reporting to Moses on his exchange with Spellman, Potter wrote in a memo, "Last night at the Lotus Club, Cardinal Spellman shook hands with Boland, and then grabbed me. He said, 'The lot you people have for us will cost us $168,000 each year. If we take a smaller lot, would we have to pay less money?' I said, 'Yes, but you have the prize lot of the Fair.' He said, 'You've told me what I want to know.' "[6] As a postscript, Potter noted for Moses's benefit that the price of the lot could be reduced if the ends were cut off, but those small parcels would not be rentable, resulting in a loss to the fair. Despite the prime real estate reserved for a Vatican pavilion, Cardinal Spellman remained on the fence regarding such a commitment.

As an alternative to the costly construction of a pavilion, the option of an off-site exhibition space was bandied about. Unlike the Brussels Fair, the New York World's Fair, Moses decided, would not compete with the city's already impressive display of visual and performing arts. Instead, events could be scheduled at the brand-new Lincoln Center (under construction at the time) and other venues to coincide with the fair's run.[7] St. Patrick's Cathedral and the Frick Museum were considered as possible sites for an exhibition of Vatican artwork. Roland Redmond, the president of the Metropolitan Museum of Art, offered the museum facilities as a possibility, as they could accommodate more people than the cathedral.

Moses was not enthusiastic about any of those options, but in a memo to Thomas Deegan he wrote, "Of course we shall accept cheerfully a verdict not to build a Holy See Exhibit building at the Fair and to use the existing facilities near the Cathedral," but he predicted that it could be regarded as a "private sectarian show." He continued,

> On further reflection, I'm not sure the analogy with the Lincoln Square Center of the Performing Arts is a true one, and fear there are possibilities of repercussions these days of emotional church and state arguments over schools, education, etc.
>
> On the other hand, I need not emphasize my own enormous respect for the judgment of His Eminence and would not think of questioning it if he has made up his mind.[8]

Moses was referring to the US Supreme Court's recent prohibition of official prayer in public schools; but as the fair was a private corporation, Moses had no such separation qualms and decided to waive the rental fee for all religious buildings.[9] Figuring into his decision was most probably Spellman's reluctance to build a pavilion, but also the (correct) anticipation that religion would be a draw. With this news, the papal envoy to the United States, Archbishop Egidio Vagnozzi, was elated.

When the pope had appointed Vagnozzi as apostolic delegate of the Vatican to Washington, DC, in 1958, the papal representative already had twenty-eight years of experience as a Vatican diplomat. If the United States had had diplomatic ties with the Holy See at the time, as it does today, Vagnozzi's title would have been nuncio, with the rank of ambassador. When Thomas Deegan, chairman of the fair's executive committee, officially called on him (with Spellman's "concurrence"), Archbishop Vagnozzi expressed his belief that "the Vatican should have one of the most memorable and effective pavilions at the Fair, however small or large."[10] Vagnozzi suggested that Moses "counsel with Cardinal Spellman about the Church setting up a small national committee made up of key figures in the Church and the laity to see how the pavilion can be brought about on a self-liquidating basis and what its general concept should be."[11]

Over the next few months, Moses and Deegan would have four visits with Vagnozzi and at least a half dozen with Cardinal Spellman to obtain a firm commitment to participate in the fair. The question was whether or not the American bishops were willing to contribute their pro rata share toward the support of the construction and maintenance of a pavilion for two seasons, which at that time was estimated at $2 million.

Spellman could not give an answer until he met with the bishops at their annual conference held in Washington, DC.[12] At this November conference, Spellman laid out his proposal for the formation of a national committee of top influential laymen and church officials to lead the effort, which the bishops approved. Almost one year later, Spellman signed the lease between the Holy See and the New York World's Fair Corporation.[13]

Spellman was not opposed to the Vatican participating in the fair, but with far more pressing needs in the archdiocese, he was strongly against spending large sums on an exhibition when money was needed to support the church's schools and hospitals. He also knew that bishops across the country would be reluctant to commit funds to the construction and operation of a Vatican pavilion without an assurance it would be successful.[14] Spellman voiced doubts that the Vatican would even lend its important works of art, as the fair would coincide with Rome's tourist high season. Undeterred, Moses consulted Roland L. Redmond, the president of the Metropolitan Museum of Art, to consider which Vatican masterpieces should be on their short list of requests.

A major cultural force in the city, Redmond oversaw the Metropolitan Museum's massive building expansion as well as an astounding doubling in visitor attendance from the start of his presidency in 1947 to its end in 1964. The Metropolitan had up to this time focused its collecting on older, "classic" art, and it was not until 1949 that the museum established its Department of American Art. To acquire and exhibit contemporary American art, the museum planned a series of competitive national exhibitions, beginning in 1950. When the invitations were sent for the first competition, eighteen painters, including Mark Rothko and Jackson Pollock, and ten sculptors, later referred to as the "Irascibles," signed an open letter to announce to Redmond that they would boycott the competition because of the jurists' "notorious hostility to advanced art." That Redmond himself was perceived as hostile to contemporary art mattered little; indeed, precisely because of this, Moses trusted his artistic judgment completely.

Redmond first suggested that the Vatican pavilion be a replica of the Sistine Chapel, "with frescoes, ceiling, etc. blown up by the best kind of photography."[15] Redmond had also assured Moses that there was practically no danger in transporting artworks for exhibitions—museums did it all the time. The museum's director, James Rorimer, had served as a monuments and fine arts officer—a "Monuments Man"—so Redmond spoke with confidence. Under Rorimer's tenure, the museum had hosted a number of important postwar loan exhibitions from European museums, among them paintings from Berlin museums in 1948, art treasures from Vienna in 1949, and Japanese paintings and sculptures in 1953. The most popular, still to come, was the two-month loan show of the *Mona Lisa* in late 1962.[16]

Over the course of the next year, various parties advanced suggestions for Redmond's short list of Vatican art treasures. One was for Raphael's *Transfiguration* altarpiece, the last work completed by the High Renaissance master. This was one of the pieces Napoleon's army had confiscated for display at the Louvre. To Napoleon, Raphael was simply the greatest of Italian artists, and the *Transfiguration* his greatest work. Redmond hesitated because it was a tempera painting on wood panel, and any change of atmospheric condition could cause serious damage, even if it were displayed for only a few months in an air-conditioned building.

Another suggestion was the *Laocoön*, one of the most famous classical sculptures. Discovered in 1506 on the Esquiline Hill, on the site of Nero's Domus Aurea, the famous sculptural group is by the Rhodian sculptors Hagesander, Polydorus, and Athenodorus and probably dates from the last century BC or the first century AD. This piece and other ancient marbles formed the nucleus of the nascent Vatican Museum collection and remain on display today. The life-size figural group is a depiction of human agony as the Trojan priest Laocoön and his sons Antiphantes and Thymbraeus struggle against an attack by sea serpents. The depiction of this life-death conflict and its exemplary display of musculature would influence an entire generation of High Renaissance artists. In 1960 the statue was restored to its original condition by first disassembling it and removing some sixteenth-century additions. Redmond believed that "the piece is so famous and the reconstruction is so new that [he was] sure it would attract a great deal of attention."[17] He was confident that it could be shipped safely, since it had just recently been disassembled and could easily be taken apart again with the pieces packed in several cases. The problem, as Redmond saw it, was that the *Laocoön*, as great as it was, might not be sufficient to attract the attention that the Vatican pavilion deserved. Even so, an additional piece of sculpture was still desirable to make the exhibit truly extraordinary.

Redmond also considered the *Apollo Belvedere*, which is equally famous. Part of the early Vatican collection, the seven-foot marble statue depicts the Greek god Apollo, whose arrow has just left his bow. A Roman copy of a fourth-century BC Greek bronze, it is dated around 130 AD. It was rediscovered before the end of the fifteenth century near San Pietro in Vincoli, the titular church of Cardinal Giuliano della Rovere, the future Pope Julius II. Recognized as the ideal of aesthetic perfection, the nude Apollo is sculpted in *contrapposto* or counterpoise, a more natural shift of weight. Positioned both frontally and in profile, he wears only his sandals and a cloak fastened at the shoulder like a Roman general's *paludamentum*. However, as magnificent and stunningly beautiful the sculpture is, like the *Laocoön* it lacked the appropriate religious value and was therefore an awkward fit for the exhibition.

Ultimately, there was only one piece that combined drawing power with a suitably religious theme: Michelangelo's *Pietà*, which, Redmond noted, was "one of the most moving and dramatic statues ever made and represents one of the finest examples of Italy's greatest sculptor." Since there were no Michelangelo sculptures in the United States, its uniqueness would add to the attraction; but more important, it carried a deeply religious theme that would be most appropriate for the Vatican pavilion. Redmond already pictured the sculpture staged alone, against a neutral background, with proper lighting, quite unlike the Baroque chapel in which it had been placed centuries ago.[18]

It could only have been the incredible audaciousness of this proposal that won over Cardinal Spellman. Like Moses, who was first to grasp the bigger picture, Spellman understood the weight of selecting the *Pietà*, regarded by many as Western civilization's finest and most beloved work of art. Now the deal took form—Spellman would raise the needed funds on the condition the pope approved the sculpture's loan.[19] On March 29, 1962, the *New York Times* broke the news: the *Pietà* was coming to the fair.

Under the headline "Fair Sees Pieta as Top Feature," journalist Gay Talese credited Cardinal Spellman for his eighteen months of tireless efforts to make this gesture possible; yet exactly how this loan was initiated remains clouded. One account credited the well-connected Gertrude Algase, who was said to have "ardently" promoted the idea to Spellman long before the loan was made public. As Spellman's literary agent, she certainly would have had his ear.[20] Spellman himself, however, turned all the credit over to Pope John XXIII, who was "delighted to do it" for two reasons. First, it would offer the chance to see the rare work for millions of people who otherwise would never have the opportunity to make a trip to Rome. And it provided the pope a vehicle to convey his gratitude to the United States for its generosity to the world's poor.[21] In reality, there would have been no Vatican pavilion if there was no *Pietà*—that was the deal.

The newsprint was barely dry when cultural elites across Europe and the United States voiced their opposition. As soon as the news broke, the transatlantic uproar began. The first reaction was one of shock—utter shock. "There must be some mistake," one European art historian gasped.[22] In Italy, a group of outraged Florentine artists organized a protest by loading a truck carrying a plaster cast of the *Pietà* and driving it through the streets of Florence with signs announcing they had removed the art treasure to safeguard it from being sent to the World's Fair. Their destination was the city cathedral, where they asked viewers to sign a petition to reverse the decision. Thinking the real *Pietà* had actually been taken hostage by art radicals, viewers soon came to understand that the sculpture was merely a plaster cast. The "art bandits" had pulled off a similar scheme six years earlier to protest the loan of forty famous paintings for exhibition at the National Gallery in Washington and the Metropolitan Museum of Art in New York.[23] One of those paintings was Botticelli's *Birth of Venus*, which was to have been sent over on an American military sea transport ship. Four artists barricaded themselves in the tower of the thirteenth-century town hall and showered leaflets into the square, calling on citizens of Florence "to rally in defense of their artistic heritage."[24] This protest had had its intended effect, and the government had canceled the tour of paintings. This time, however, the protest was in vain because the Italian government was not the owner of the *Pietà*—the Vatican was.

On April 3, *Il Giornale d'Italia* ran an interview with Redmond, who defended the decision. Since the end of World War II, he noted, the Metropolitan Museum had been responsible for transporting Ivan Meštrović's five-ton *Pietà* from Yugoslavia. The museum had already shipped the *Rondanini Pietà*, Michelangelo's unfinished marble sculpture, from Italy to New York and back. Even before the war, Italy had sent over

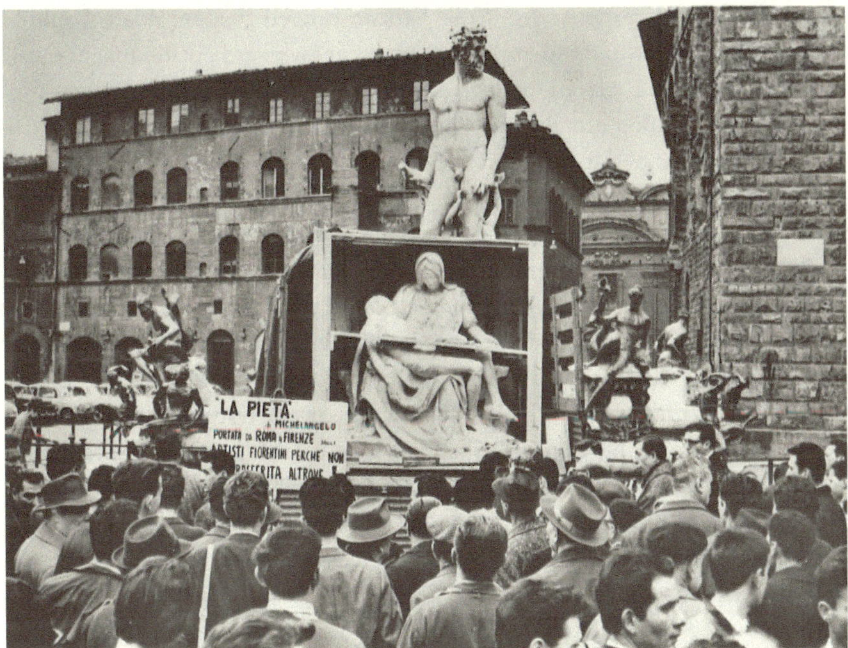

Figure 3.1. In Florence, artists display a replica of the *Pietà* to protest the shipment of the statue to New York. (UPI wire photo)

Michelangelo's *Virgin with Child*, Donatello's *Saint John*, and Botticelli's *Primavera* from the Bargello Museum for the San Francisco International Exposition in 1939.[25] At the University of Rome, however, art history professor Giulio Carlo Argan, who later became the first Communist mayor of Rome, was so startled by the news that he described the move as "the most absurd that can be imagined." He said that "the best homage the Americans could pay to the great man in the fourth centenary of his death would be to renounce this foolish idea."[26]

Adding to the objections was that "this very risky act of generosity to the Americans" came at a particularly inopportune time, when the country was preparing to celebrate the fourth centenary of Michelangelo's death in 1964. The popular Italian newspaper *Il Messaggero* scornfully noted that the Americans "would like to see the Pietà lined up with their most advanced industrial products."[27]

The limited response from the Vatican offered that the loan would present an opportunity to improve the present awkward positioning of the statue, which was placed too high and poorly lighted. The official newspaper of the London archdiocese, the *Tablet*, noted ironically, "It is hardly necessary to move it to a better position via New York."[28] The paper reported that the decision was made by lower levels of the Vatican. Stung by the criticism, Spellman, while speaking at a commencement ceremony in which Cardinal Heenan, an English prelate, was receiving an honorary degree, reminded the Englishman that at least the *Pietà* loan was a two-way journey—a veiled reference to the Elgin Marbles that were taken from Greece's Parthenon and are still in London two centuries later.[29]

Il Messaggero placed the blame squarely on the pope, criticizing him for accommodating Cardinal Spellman's request. The paper wrote that the pope "had shown the weakness of a father who cannot say no to his children who ask him to entrust things to them that cannot be entrusted to them."[30] Spellman had been in Rome, attending a preparatory meeting for the upcoming Ecumenical Council (Vatican II), and by the time he returned home, letters of protest began appearing in all the major American press outlets. Germain Seligman, an art historian and gallery owner who had been a member of the Art Commission for the 1939 New York World's Fair, wrote a letter to the *New York Times* stating that the 1939 commission had considered requesting such works of art for exhibition but considered the risk too great and consequently abandoned the idea. Responding to the assurance given by the fair organizers that the *Pietà* would be insured for $5 million, Seligman wrote that insurance coverage was mere "sophistry, as no millions of dollars could ever replace the Pietà."[31]

The journal *ARTnews* lost no time in referring to the announcement as the "Michelangelo Scandal" and likened the move to pillage. In an editorial, Alfred Frankfurter wrote that moving the *Pietà* amounted to "the most drastic and gratuitous act of vandalism Rome will have seen since the fifth-century barbaric invaders from beyond the Vistula gave their name, *Vandali*, to willful destruction."[32] Believing the *Pietà* to be "as fragile a complete work of art as exists on the face of the earth," Frankfurter noted that the sculpture would have to be lifted by derrick or crane a total of four times, equaling four opportunities for disaster during the first half of the round-trip passage alone. The first lifting would be at the point of origin, then to deliver it to the ship or airplane, again at landing at dock, and, finally, at "Mr. Moses' bazaar on Flushing Meadows." Frankfurter also highlighted the possibility of internal fissures or imperfections in the sculpture, which could spread if it was simply moved, much less transported to another nation.

He dismissed the fair as a "nine-days wonder of a kermess"—a kind of fun fair—and that there were "even beggars too proud to accept loans with dishonor attached." He further appealed to Cardinal Spellman, with his "superb war record as a member of the famous Roberts Commission," to reconsider the move.[33] Reading of the reaction to the "artistic coup," as the *New York Times* described it on April 11, the former assistant secretary of war John McCloy immediately penned a letter to Spellman to thank him for his efforts and wrote that "instead of being despoilers as most conquerors have been in the past, the Americans were really great preservers of art." From McCloy's point of view, the loan was "the least they [the Italians] could do."[34]

McCloy publicly thanked and defended both the cardinal and the pope in a letter to the *New York Times*, stating that the loan was justified, considering the extraordinary contribution that the United States made during World War II to preserve European art and antiquities, especially in Italy. Included in this were Michelangelo's *David* and his *Deposition* (also known as the *Florence Pietà*), which the Monuments Men had retrieved from the crypt of the Duomo in Florence during the summer of 1945. As an officer in the army, McCloy was present when the works were unwrapped from their burlap and straw wrappings and examined with only the illumination of a single lightbulb. McCloy added that such protective measures were accomplished at considerable risk to the lives of the soldiers. Hundreds of thousands, if not millions, of dollars were spent to restore and preserve these great works, along with buildings such as Milan's Santa Maria delle Grazie convent, which houses Leonardo da Vinci's *Last Supper*.[35]

Rebutting McCloy was the chairman of the Department of Fine Art at the University of Pennsylvania, Frederick Hartt, who was one of the Monuments and Fine Arts officers who accompanied McCloy on his tour of Florence. Hartt recalls how he, along with thirty-two former officers, signed a letter to protest the shipment of paintings from Germany to the United States with "ironclad guarantees of security." Unfortunately, one of the paintings in the group, Domenico Veneziano's tondo *Adoration of the Magi*, suffered an irreparable vertical split along almost the full length of the work's diameter. Questioning whether the loan of the *Pietà* was deserved, Hartt felt that the United States had done what it needed to do not to elicit gratitude but to protect these works of art for generations to follow. That "to expose them to possible injury now is only to renew the very dangers which the monuments and fine arts officers of England and America risked their personal safety to prevent."[36]

As the date of transport neared, more art world elites weighed in. On May 25 the *New York Times* printed a letter by John Coolidge, director of the Fogg Art Museum at Harvard University, who was appalled by the idea of shipping a revered masterpiece to the fair. He asked, "What right have we to remove this work of art, even for a brief period of time, from its intended and rightful location, a location where it is viewed every day by hundreds of pilgrims from all over the world?"[37]

Considering the setting inappropriate at best, and offensive at worst, Coolidge introduced a new objection of juxtaposing piety with commercialism, in addition to the biggest concern, the risk of damage. Characteristically blunt, Robert Moses responded that it was a matter for the church authorities: "The statue was made for them. It belongs to them."[38] Cardinal Spellman also weighed in, responding to criticism from museum directors who themselves "enriched their museums with holdings from abroad, and why we can't have another to enrich the fair, I don't know." Spellman also couched it as an equity issue—those who have money to travel to Italy can see the artwork for themselves, but for those who do not have the money, they could come to the fair and see it for free.[39]

Many of these same objections were repeated when the First Lady, Jacqueline Kennedy, asked the French minister of culture André Malraux, during his official visit to Washington, for a loan of the *Mona Lisa* for display in the National Gallery of Art. Edward T. Folliard, the *Washington Post*'s White House reporter, first broached the idea of a loan at a luncheon hosted by the Overseas Writers, the diplomatic reporter corps, in May 1962.[40] Open to the idea, Malraux announced the plan upon his return to Paris, without consulting museum officials first. The shocking news "fell upon the Louvre like a bomb."[41] A chorus of indignation rose from all corners of French society. Echoing the Italian criticism of the pope directed at Cardinal Spellman, the French accused Malraux of succumbing to the charms of the American First Lady. The newspaper *Le Figaro* wrote that "the unbelievable news item seems to be true" and asked the American people to refuse the loan.[42] Despite resistance, the *Mona Lisa* went on display at the National Gallery on January 8, 1963, and not quite a month later was shipped to the Metropolitan Museum of Art in New York City for display. It remained on display until it made its return voyage to France on March 7.

In advance of the *Mona Lisa*'s arrival in the United States, Moses had requested that the French government, if it would be willing, lend the portrait for display at the World's Fair. Even before Moses made the request, the idea of the *Mona Lisa* coming to the World's Fair had been floating around. A member of Jacqueline Kennedy's White House Fine Arts Committee, Mary Lasker, wife of the advertising giant Albert D. Lasker, promoted the painting's display at the French pavilion.[43] Reportedly, André Malraux was in favor of the loan, but only if it was to be displayed in a government-sponsored pavilion. Since there would be no official French presence at the fair, the French government would consent only if it was hung in the United States pavilion, rather than in an unofficial French pavilion.[44] Apparently that was not to be, and after the *Mona Lisa*'s visit to Washington and the Metropolitan Museum in New York, she returned to her home at the Louvre.

On June 3, 1963, after months of suffering, the eighty-one-year-old John XXIII died of peritonitis. Not three weeks later Giovanni Cardinal Montini was raised to the papal chair and took the name Paul VI. Born in 1897 in Concesio, a small town in the province of Brescia about sixty miles east of Milan, Montini, whose father was a lawyer, journalist, and politician, was of a less humble background than his predecessor. Though frail in health, he had been ordained in 1920 and two years later entered the Vatican diplomatic service. As he rose in rank, he became a trusted adviser to Pius XII, a crucial role during the painful years of World War II. In 1954, Montini was appointed archbishop of Milan, the largest diocese in Italy, and four years later John XXIII named him a cardinal. Considered progressive for his ecumenism and work with labor unions, he was neither politically Left or Right. As the new pontiff, he immediately inherited the remaining three sessions of the Ecumenical Council. His diplomatic career trained him to carefully assess complex issues from all perspectives, contributing to a sense of balance.[45] While Paul will probably be most remembered for his ban on artificial birth control, *Humanae Vitae* (Of human life), in the early days of his papacy he inherited a thorny problem of a much less sweeping nature.

With a change of Vatican leadership, art critics across Europe wasted no time in pressing the pope to reverse his predecessor's agreement to loan the *Pietà*. In early September, the *New York Times* reported a rumor "from high Vatican sources" that, owing to public pressure against the loan, Pope John XXIII had been reconsidering the decision to send the *Pietà* to America. The next day, Vatican officials denied any knowledge that the pope had changed his mind before his death.[46] The decision now had the finality of a papal encyclical.

While the debate to loan the *Pietà* nevertheless raged on, attention shifted over to Greece, where, on September 13, 1963, the newspaper *Kathimerini* proposed sending the fourth-century BC statue *Hermes and the Infant Dionysus* by Praxiteles to show in the New York fair. With public support actually in favor of the move, the paper quoted one reader as saying, "The American from Utah or Milwaukee who visits the Fair can hardly be expected to show interest in the chick-peas of Santorini or the prunes from Skopelos, which the Greek pavilion will display."[47] That the Greek people saw the fair as a valuable opportunity to highlight the country's rich and deep cultural history must have pleased Moses. The commissioner general of the Greek pavilion, girding for the internal battle he knew he would have to face, asked for support from American art groups to convince the Greek government to allow the sculpture to be sent over. Alfred Block, president of the National Sculpture Society, a vanguard in the defense of representational art, gave his full support to the move. He understood the ancient Greek masterpiece's potential influence on contemporary sculpture, which had evolved to complete abstraction. In a letter to the *New York Times*, he wrote, "The art world and the general public too need the stabilizing influence of such a time-tested artistic accomplishment."[48] A four-member World's Fair delegation, led by Gates Davison of the fair's International Affairs Department, flew to Greece to persuade cabinet members to allow the shipment, citing the safeguards provided for the shipment of the *Pietà*—but to no avail. Greece's Archaeological Council rejected the proposal, citing the legal prohibition against exporting antiquities and the risk of damage or theft.

In March 1964, France loaned the Louvre's most precious marble statue, the *Venus de Milo*, in honor of the Olympic Games in Tokyo. It would be only the second time the statue left Paris; the first was two decades earlier, in 1940, when it was taken to a country chateau to protect it from possible war damage.[49] Seized by France as Napoleon's war booty in 1822, the statue would pose less of a problem, French officials said, than that encountered when the *Mona Lisa* was sent to the United States. The 1.5-ton *Venus* was sent on a thirty-three-day voyage on the French liner *Vietnam*. It arrived on March 22, 1964, at the Museum of Western Art in Yokohama, where, once it was unpacked, the museum discovered that four pieces had been chipped off. Three of the chips were plaster used in restoring the statue, and the fourth was a piece of original marble located beside the statue when it was discovered in 1820 on the island of Milos.[50] Despite the fact that the damage was relatively minor, it reignited the debate on shipping the *Pietà*.

Ignoring the potential positive impact of the *Pietà*'s profound spirituality on the fair crowds, the *Christian Century*, the voice of mainstream Protestantism, seemingly more concerned about causing offense, editorialized that the loan would "appear to some Europeans as typical American arrogance." To support its disapproval, it cited the recent disaster of the Greek-owned cruise ship *Lakonia*, which had caught fire and sunk north of the Portuguese archipelago of Madeira on December 22, 1963, with the loss of 128 lives.

Realizing that nothing short of a miracle would prevent the move, Alfred Frankfurter and other critics also directed their appeal to protecting the American image abroad. Frankfurter accused the loan of prompting "the worst possible 'press' throughout the world." Almost unhinged, he railed against this "Hollywooden kind of churchmanship," and continued that this was "the most blatant vulgarity, a demonstration of unfeeling publicity-seeking that will allow those unfriendly to America to remind the world that our national taste and ethics are still on the level of P. T. Barnum and his suckers."[51] In a similar vein, G. Prezzolini of the Italian magazine *Borghese* asked, "What has the Pietà to do with a purely mercantile undertaking as vulgar, as noisy and as reeking of money as a world's fair? All you want is an American . . . to confuse Christ with the Money-changers."[52]

Two months before the fair's opening, Frankfurter acknowledged that there would be no satisfaction in saying "I told you so." He continued: "Let this be the last occasion on which the civilized world has to hold its collective breath—to find out the results of one more whim of general or cardinal or any other ruler who believes art is merely a vehicle for international advertising. After five centuries, the world owes its Michelangelos more than to send them out barnstorming as if they were so many Miss Rheingolds."[53] Frankfurt was referring to the popular Rheingold Beer–sponsored beauty contest. In response to the many critics of the loan, Moses, in his typical pugnacious manner, countered that "art pundits who had hardly been aware of the existence of the *Pietà* became its self-anointed guardians" and had the "effrontery to instruct the authorities of the church as to their responsibilities to mankind."[54]

Taking a softer tone, Rev. Clement J. McNaspy, associate editor of *America*, the Jesuit weekly magazine, wrote that "seeing the Pieta is infinitely preferable to not seeing it at all." His only reservation was that he hoped that "the [Vatican pavilion's] curators see to it that the vital religious art of today is boldly emphasized, otherwise they risk performing a real disservice—and at the very least they have missed a vast opportunity."[55] He was not alone in fearing that the pavilion would overlook contemporary art. Choosing to showcase a Renaissance sculpture seemed to some as a look backward. They didn't know it at the time, but the pavilion directors were keenly aware of the impact of contemporary religious art and would integrate the new with the old throughout the exhibits.

Americans were not used to seeing fine art outside the museum setting, especially when that setting was a fair with a carnival sideshow atmosphere. There was that element at the fair, but Moses saw to it that it would be so much more. There were other pavilions that featured fine art, and art educators recognized the opportunity. Even though "mere exposure is not enough," it was a beginning, and they applauded the efforts, as long as they were being done for the "right motives."[56]

Never had the *Pietà*, the only sculpture that bears Michelangelo's signature, been removed from St. Peter's Basilica. This was true, but it was not the first time the sculpture had been moved.

4

The Coup

Moses Gets His Blockbuster

The name of Michelangelo's Renaissance masterpiece, *Pietà*, means "pity" or "sorrow." The name is appropriate, as the sculpture depicts the grieving Virgin Mary holding the dead body of her son, Jesus, in her arms. From the moment the work was unveiled in 1499 to the present, it has been recognized as one of the most compelling sculptures ever made, the ultimate fusion of the Florentine's creative genius and religious belief.[1] So iconic is the work that the word *Pietà* evokes only one image—Michelangelo's.

Michelangelo Buonarroti was born on March 9, 1475, the second of five sons, to a family descended from the counts of Canossa in Tuscany. His parents, Ludovico and Francesca Buonarroti, lived in the small village of Caprese, near Arezzo and south of Florence, where Ludovico served as *podestà* or mayor. The family moved to Florence soon after Michelangelo's birth, but his mother, not strong enough to nurse the infant, took him to the nearby town of Settignano to be nursed by the wife of a stonecutter, who was also the daughter of a stonecutter. Michelangelo later jokingly attributed his skill as a sculptor to the fact he had been wet-nursed by a stonecutter's daughter.[2]

As a young boy, Michelangelo began his instruction in grammar and writing, but he preferred drawing to studying. He spent hours sketching the fresco paintings by Masaccio in the Brancacci Chapel, which is located in the Church of the Carmine in Florence. At the age of thirteen, he was apprenticed to the *bottega*, or studio, of Domenico Ghirlandaio for three years. By one account, when Michelangelo with his friend Francesco Granacci went to see the collection of classical sculpture in the Medici gardens in Florence, he was so overwhelmed that he never returned to Ghirlandaio's studio. The sixteenth-century biographer of Renaissance artists and a contemporary of Michelangelo, Giorgio Vasari, gives another account. Lorenzo "the Magnificent" Medici, as a patron of arts, had established a school in 1489 to promote the study of art. He charged Bertoldo di Giovanni, a student of the fifteenth-century sculptor Donatello, with supervising his sculpture garden and mentoring young sculptors. Lorenzo asked Ghirlandaio and other masters to refer their most promising apprentices to his school, and it was at this time that Ghirlandaio sent him both Michelangelo and Granacci.

When Lorenzo saw the young Michelangelo working on a copy of a marble mask of a faun, he realized Michelangelo had a special talent and asked Michelangelo's father for his consent to allow his son to live and learn under Lorenzo's guardianship. With a room at the Palazzo de' Medici and a monthly pay of five ducats, Michelangelo lived as a Medici family member for the next three years. Also living in the palace were some of the most brilliant Florentine poets, artists, musicians, and thinkers, including the philosopher Giovanni Pico della Mirandola, who influenced the young artist.[3]

During this period, at the age of sixteen, Michelangelo sculpted his first depiction of the Madonna and child, the *Madonna of the Stairs* (ca. 1491), a marble bas-relief in which the Virgin is seated in profile at the foot of a flight of stairs nursing the Christ child. Michelangelo returned to his father's home in 1492 when Lorenzo died but was invited to return by Piero de' Medici, Lorenzo's eldest son and successor. Piero's rule of Florence was cut short after only two years when the Florentines expelled him and his family on November 9, 1494, for betraying the city to the French king Charles VIII. Michelangelo had already left Florence for Venice on the advice of his friend, Andrea Cardiere, another guest of Piero, who had a premonition that the Medicis would be ousted. After living first in Venice, then Bologna, Michelangelo returned to Florence in 1495, only to leave for Rome one year later. There, he began work on the life-size figure of *Bacchus*, a male nude of the pagan god of wine, acquired by the Roman nobleman and banker Jacopo Gallo, who became his patron and agent. It was through Gallo that Michelangelo received his contract for the *Pietà*.

The *Pietà* was commissioned in 1498 by Cardinal Jean Bilhères de Lagraulas. Lagraulas has often been eclipsed by the artwork he commissioned, but he was an extraordinary man in his own right. At a young age, he entered the Order of Saint Benedict and in time was ordained a priest. He rose steadily through the ranks and was appointed abbot of the Royal Abbey of Saint-Denis in 1474, which held a privileged status as the burial site for the kings of France. Skilled as a diplomat, Lagraulas became Charles VIII's ambassador to the Holy Roman Empire and in 1491 was appointed to the role of French ambassador to the papal court. Two years later, he "received the red hat" as a cardinal priest of Santa Sabina, the historic church on Rome's Aventine Hill, after having faithfully served kings Louis XI and Charles VIII in Spain, Germany, and at Rome.[4] As a remembrance of his service to France and the Holy See, Lagraulas requested and was granted permission by Pope Alexander VI to place the sculpture in the Chapel of Saint Petronilla, the French chapel of the kings in the old St. Peter's, where he arranged to be buried.[5]

Cardinal Lagraulas signed the contract with Michelangelo's agent, Jacopo Gallo, on August 27, 1498, making Gallo the guarantor for both the cardinal and the sculptor, in addition to detailing the terms and conditions. The contract reads as follows:

Rome, 27 August 1498

Grant to Michelangelo for a marble group of the *Pietà* in Rome.

On the 27th day of the month of August 1498

Be it known and manifest to all who will read this document that the most Reverend Cardinal of Saint-Denis has agreed with master *Michelangelo*, Florentine sculptor, that the said master will make at his [the Cardinal's] expense, a *Pietà* in marble, that is a Virgin Mary clothed, with

Christ dead in her arms, as large as a [life-size] well-proportioned man, for a price of 450 gold ducats of papal coinage, to be paid within one year from the beginning of the work. And the said Most Reverend Cardinal promises [agrees] to pay in the following manner, that is, to start with, he promises to pay 150 gold ducats of papal coinage before the work is started, and once the work is started he agrees to pay 100 ducats of the same coinage to the said *Michelangelo* every four months, so that the aforementioned 450 gold ducats of papal coinage will be completely paid in one year, if the said work will be completed; and if the work will be completed before the year [is over], his Most Reverend Lordship will pay the remainder [balance] at once.

And, I, Iacopo Gallo, guarantee [promise] the Most Reverend Monsignore that the said *Michelangelo* will complete the said work within one year and that it will be the most beautiful work in marble to be seen today in Rome, and that no other master could produce a better work. And, on the other hand, I guarantee [promise] the said *Michelangelo* that the Most Reverend Cardinal will pay in accordance with the above stipulation. And in witness thereof, I, Iacopo Gallo, have written the present document with my own hand, on the above inscribed year, month and day. It is understood that this document renders null and void any other such document written by me or by the hand of the abovementioned master Michelangelo, and that the agreement is the only valid one.

The Most Reverend Cardinal has given to me, Iacopo, some time ago, 200 gold ducats of Chamber coinage and today 50 gold ducats of papal coinage.

It is so Ioannes, Cardinal Dyonisii

I agree to the same Iocobus Gallus,

By his own hand.

The subject of Michelangelo's sculpture was not new. On the contrary, the image of a mourning Mary with the body of Christ crucified dates back to Byzantium. The *Pietà* motif is a variation of the Madonna-and-child theme, which is prevalent throughout all Christian traditions. The earlier image of the Madonna seated with the Christ child on her lap evolved into the grieving mother who holds her dead son on her knees.

In images and sculptures of the Madonna and child, the Christ child sits, stands, or lies across the Madonna's lap. Generally, the Madonna sits on a throne, robed, static with her knees and feet close together.

The *Pietà* motif evolved in France and Germany in the form of Gothic sculpture, though there is no direct scriptural reference in the Passion of Christ narrative itself. Because of this, scholars describe the poignant image of the *Pietà* as a nonhistorical *Andachtsbild*, or devotional image, that unites the viewer with Mary in her suffering, in fact with the suffering of all humanity.

Prior to the fifteenth century, the *Pietà* had not yet caught on in Italy, but the mystics Bernardo da Siena and Suso of Germany popularized conceptions of the subject in Florence by the end of the century. As its popularity spread, the theme took on an emotional intensity in devotional images.[6] Representations of the body of Christ stretched horizontally across the Virgin's knees, with Saint John and Mary Magdalene on either side supporting his head and feet, were completed by a number of Renaissance artists, such as Perugino, Giovanni della Robbia, and the school of Ghirlandaio. Placed upon the altar, the image of Christ's body evokes the Eucharistic sacrifice.[7] Therefore, Michelangelo sculpted the group with an emphasis on the dead body of Christ, rather than on Mary. Here on Calvary, Mary, heavily draped, sits on a rock, her left hand is slightly tilted open, not touching Christ's body. Her right hand supports him, but buffered by the shroud. Christ's body rests on Mary's lap as the wafer of the host rests on the white corporal. The Virgin's lap becomes the altar on which the sacrifice is laid. Mary is the "living counterpart of the altar-table," on which the "bread of life" rests during Mass.[8]

Michelangelo sculpted the marble *Pietà* in the round, with mother and son forming a pyramidal group. Michelangelo refers back to the older type of the seated Madonna, giving the natural impression that the Virgin holds her child in her arms. She bows her head as a model of heroic motherhood, a Renaissance ideal.[9] She does not weep, as weeping in sculpture was not heroic. Michelangelo altered her natural proportions by extending the knees of the Virgin, as if Mary herself becomes the throne on which Christ's body rests.

In earlier such works, the Virgin was shown as older, but Michelangelo depicts her as a young woman. This prompted "pointless speculation" during Michelangelo's lifetime about the discrepancy between Mary's depiction and what would have been her actual age.[10] If one assumes her age to be eighteen at the time of Christ's birth, thirty-three years later, at the time of Jesus's death, Mary would have been fifty-one. In response, Michelangelo remarked that "a pure virgin will retain the appearance of youth much longer than a married woman."[11] This was not a new concept, but one already echoed in Dante's *Paradiso*: "Virgin Mother, daughter of thy Son," expressing that Mary was the model of the church. The painter Sandro Botticelli included this passage on the steps to Mary's throne in his painting *Enthroned Madonna with Saints* (mid-1480s). When the Florentine condottiero and banker Giovanni Battista Strozzi (the Elder) composed a poem for the installation of a replica of the *Pietà* in the church of Santo Spirito in Florence, he recalled Dante's passage:

> He is also, in spite of Himself,
> Our Lord and they
> Spouse, son and father,
> O His only spouse, daughter and mother.[12]

These expressions and Michelangelo's *Pietà* echo the doctrine of the Roman Catholic Church in which the Virgin is shown as "the mortal vessel of Divine Grace, the body through which divinity took on man's flesh."[13] Art historian and "Monuments Man" Frederick Hartt succinctly states that the question of age is "irrelevant" and dismisses any attempt to literally translate the work, since Michelangelo was concerned with the deeper spiritual meaning.[14]

The body of Christ reflects Michelangelo's profound grasp of anatomy.[15] He depicted Christ as slender, in contrast to the artist's hyper-musculature in the sculpture *The Battle of Lapiths and Centaurs* (1490–1492) or the sensuality of his *Bacchus*, as neither body type was appropriate for this subject. The art historian Dr. Josef Vincent Lombardo defines Neoplatonism as the "glorification of beauty [that] was a manifestation of the divine. . . . The morality and dignity of man were an inheritance from God to enable man to ennoble his life on earth." Michelangelo considered Christianity and Neoplatonism compatible and valid expressions of universal truth.[16]

The contract between Michelangelo and his patron stipulated that the sculpture was to be "the most beautiful work in marble to be seen today in Rome, and that no other master could produce a better work." The cardinal was rewarded. Vasari wrote in his *Lives of the Artists*, "No sculptor, not even the most rare artist, could ever reach this level of design and grace, nor could he, even with hard work, ever finish, polish, and cut the marble as skillfully as Michelangelo did here, for in this statue all of the worth and power of sculpture is revealed."[17]

The Chapel of Saint Petronilla where the *Pietà* was to be installed was originally built as a mausoleum for the emperor Constantine's half-sister, Anastasia, and was referred to as the Honorian Mausoleum; it was directly attached and aligned with the south transept of the basilica. It was later dedicated to Saint Petronilla, who was believed, incorrectly, to be the daughter of Saint Peter himself. Her body was reinterred or "translated" to the mausoleum from the catacombs by Pope Stephen II in the middle of the eighth century.[18]

From the time of Charlemagne, who visited the chapel and donated funds for its decoration, the chapel's upkeep and further embellishment were financed by the French kings. The rotunda was known as the Capella Regum Francorum or, in Italian, Cappella del Re di Francia, and was the focus of French piety in Rome and a symbol of France's close ties with the Holy See. For centuries popes had guaranteed the French king "the right to have masses said and to appoint or approve the appointment of the chaplain or chaplains who officiated at the altar."[19]

Michelangelo was only twenty-five years old when he completed his entirely original interpretation of the *Pietà*. Vasari used an anecdote to explain why the artist felt compelled to somewhat inappropriately add his name to a sculpture for the first time: Michelangelo had supposedly returned to view his sculpture and found a number of pilgrims from Lombardy praising the work. He overheard one ask another who the sculptor was and heard the reply, "Our Gobbo from Milan" and was stunned that his work was being attributed to someone else. He resolved to right this wrong by returning at night with his lamp and chisels, locking the doors, and carving in the most beautiful Renaissance lettering on the strap that crosses Mary's heart: "MICHAELA[N]GELUS BONAROTUS FLORENTIN[US] FACIEBA[T]"—Michelangelo Buonarroti, Florentine, was making this.[20]

The cardinal did not win his race with time and died in Rome before the sculpture was installed in the chapel, although we want to believe that he may have viewed the work in progress while in Michelangelo's studio workshop. It stood in its originally intended chapel for only a couple of years before it began its mini-tour of the basilica, one step ahead of Pope Julius II's wrecking ball. The Chapel of Saint Petronilla was demolished in the first quarter of the sixteenth century to make way for the foundation of the new St. Peter's.

After the chapel's demolition, the *Pietà* was transferred to the adjacent Cappella della Vergine Maria della Febbre, a setting that matched its original location as closely as possible. This rotunda chapel was also built as a mausoleum but dates to the pre-Constantinian Roman circus, the open-field chariot-racing stadium and mass entertainment venue that once occupied the site and was believed to be where Saint Peter was crucified. The archaeologist Ferdinando Castegnoli investigated the buried foundations of Santa Maria della Febbre in 1957 and found stamped bricks dating the structure to the reign of the emperor Caracalla, from 198 to 217 AD. As the foundation was filled in with rubble to level the site for the basilica, it is unknown who exactly was buried here, but most likely it was a member or close associate of the imperial family. The remaining structure standing aboveground was rebuilt, to mirror the new Honorian Mausoleum built directly to the west and was used as the basilica's sacristy.[21]

In its second move, the *Pietà* was temporarily transferred to the new sacristy, the Chapel of the Choir, in 1568, four years after Michelangelo's death. Cardinal Antonio Carafa covered the cost of the move.[22] The chapel was where the canons of the basilica celebrated daily the Liturgy of the Hours as a choir. But when that chapel, too, was demolished in 1609, and when the canons temporarily took possession of the altar of Saints Simon and Jude in the newly built basilica, they also moved with them the *Pietà*, which they now considered theirs, and placed it over the altar. Once their new chapel was completed, they were allowed to again take the *Pietà* with them, but not without a fight. The dispute escalated and had to be settled by the pope, in 1625.[23]

In 1626, the Baroque sculptor Francesco Borromini designed an oval pedestal of red marble bordered with a white cornice, on which the *Pietà* was placed. In 1637, Count Alessandro Sforza di Piacenza presented two bronze putti holding a bronze crown to hover above the head of the Virgin, following a devotional practice of crowning images of Mary regarded as miraculous. Though there is no record of a miracle, the creation of the *Pietà* itself was considered the miracle.[24]

Nearly a century later, in 1749, the *Pietà* was shifted to its present location, the first chapel on the north side of St. Peter's, the Chapel of the Crucifix, now renamed Cappella della *Pietà*. Behind the figures, against a background of multicolored marble bordered in black, a cross of white marble was set, a reference to Golgotha, the site of the Crucifixion. This backdrop was a newer addition and not part of the original plan. Here the *Pietà* would remain undisturbed for more than two hundred years.

In 1961, a new, lower pedestal was installed, more in line with Michelangelo's original intention. As it had been displayed, the sculpture was too high to be viewed properly by visitors. This had been known for some time; the nineteenth-century Swiss art historian Heinrich Wöfflin remarked that "it is most barbarously placed—raised so high that it is impossible to get the chief point of view—lost in vast space."[25] The change was apparently suggested by Professor Redig de Compos, the head of the Pontifical Museum, Gallery, and Monument Commission at the time. He was anxious to have the angle at which the statue sat on its pedestal changed to the position that Michelangelo originally intended. This repositioning was led by Francesco Vacchini, head of the St. Peter's operation and maintenance corps, the "Sanpietrini."[26]

With the critical reception and praise he garnered for the *Pietà*, Michelangelo's stature rose, both in Florence, the heart of the Renaissance, and Rome, where he would become the most sought-after artist by the papacy. Since he viewed himself primarily as a sculptor, having completed his much-celebrated *David* sculpture for the city of Florence, he hesitated to take on Julius II's request in 1508 to paint the ceiling of the Sistine Chapel. However, his originality and mastery of the human body both in paint and marble guaranteed his prominence throughout the remainder of his lengthy professional career—long enough to serve nine popes. In addition, his architectural plans were so innovative that he established new design standards in Florence and Rome. His Laurentian Library and the New Sacristy for the Medici family established him as an architectural authority, leading to his appointment as papal architect for the rebuilding of the new St. Peter's. Michelangelo mastered all three of the visual arts: sculpture, painting, and architecture, and earned the Renaissance appellation of "universal genius."[27]

Later in life, Michelangelo sculpted two more versions of a *Pietà*. In 1547, he began a grouping of Mary, Christ, Mary Magdalene, and Nicodemus or possibly Joseph of Arimathea—both of the latter were at Golgotha, the site of the Crucifixion. The face of the Joseph/Nicodemus figure is said to be Michelangelo's own. Michelangelo was so displeased with his own work that he gave it to a friend, Francesco Bardini, but not before he took a hammer to it. Bardini, in turn, commissioned another sculptor, Tiberio Calgagni, a friend of Michelangelo's, to continue the work, which remains unfinished. The work, traditionally referred to as the *Deposition* or the *Florentine Pietà*, can be viewed at the Duomo Museum in Florence.[28]

The *Rondanini Pietà* was Michelangelo's last work, which was also left unfinished. With elongated figures, Mary and Jesus are depicted vertically, with Mary holding her son's body as it collapses to the ground. Michelangelo had begun the work in 1552 and was working on it until his death in 1564. It takes its name from the Palazzo Rondanini, where it stood, and is now on view at the Castello Sforzesco in Milan.

But the most transcendent (and the only finished) of all the *Pietà*s was that in St. Peter's, and it would soon embark on its voyage across the Atlantic Ocean.

5

Spellman's Dream Team

The Men Who Made It Happen

Participating in a world's fair was a new enterprise for Spellman, but he was fortunate to call on the experience of Howard S. Cullman, who had served as the US commissioner general for the Brussels World Fair of 1958. President Eisenhower had appointed Cullman as head of the US pavilion with the rank of ambassador, and, like Robert Moses, Cullman was a power broker in New York City, albeit a much more affable one. Born into a wealthy, tobacco-growing family, he took charge of the family business after the death of his father. In addition to his commitments as a tobacco, real estate, and investment tycoon, he served as commissioner of the New York Port Authority for forty-two years and chairman for ten. Involved in public service and philanthropy, he was regarded as a Broadway "angel" who invested in theatrical productions.[1]

Cullman was also a member of St. Patrick's Cathedral church and made himself available as a consultant to the archdiocese during the early months of pavilion planning. He offered to open the books of the US pavilion in Brussels to an ad hoc committee to assess the cost and effort involved in the operation of a pavilion. In November 1960 he wrote to Monsignor Charles J. McManus, a priest serving at the cathedral, "To my way of thinking you have a very different problem [as opposed to a national pavilion]. You will not have industrial shows, Vogue models, Circarama, Brass Rail operations [food service operators] and many concomitants that go with the average exhibition of a nation showing its way of life."[2] With a broad knowledge of European arts and culture, Cullman wrote that "the Catholic Church has a lot to offer a suggested attendance of approximately 75 million" and recommended an auditorium be considered, in which Gregorian chant and instrumental and other choral music could be featured. He suggested "that the great art of the world, originating in Italy . . . be available for such an exhibit."[3]

This suggestion set the course for what would be the fair's biggest attraction. On the more practical, day-to-day management, Cullman recommended a "lend-lease" arrangement with universities and seminaries to build up a corps of volunteers, similar to the hospital system in New York, where nuns contributed their services for room and board. Cullman also mailed a carbon copy of his letter to Cardinal Spellman.

Initially, the Archdiocese of New York was reluctant to publicize its commitment, as its directors had not yet determined how they would finance the pavilion. It was possible that Spellman might want to bring this matter before the Second Ecumenical Council of the Vatican the coming October.[4] But whether or not he had the chance to broach the topic, the archdiocese decided to raise the funds completely in the United States.

In a private audience with the pope in October, Spellman presented the design concept for the Vatican pavilion. The total cost of the pavilion was estimated to be $3.8 million[5]—this included construction, maintenance, programming, the exhibits, and the final demolition and removal from the site, an obligation of all exhibitors at the end of the fair. For a building that would stand for only two years, this was a sum immensely higher than what the church would normally spend. But considering that the General Motors exhibit—the only pavilion that would receive more visitors than the Vatican's—was estimated at $50 million, it was a rather modest sum.[6]

The planning and construction of the pavilion exhibit was to be a coordinated effort between Rome and New York City. However, the responsibility for designing and building the pavilion fell squarely on Cardinal

Spellman and Bishop Bryan J. McEntegart, whose diocese encompassed the Flushing Meadows fairground.

McEntegart had been head of the Diocese of Brooklyn (which covered the boroughs of both Brooklyn and Queens) since 1957. The diocese had been established in 1853 from territory of the Archdiocese of New York, at a time when Brooklyn was still a separate city from New York. The most pressing needs at the time were the care of millions of Irish Catholic immigrants, many of whom settled in Brooklyn and Queens. Originally including all of Long Island, Nassau, and Suffolk Counties, the territory was divided ten days before Bishop McEntegart was installed, to form the newly established Diocese of Rockville Centre. In area, the Diocese of Brooklyn is the smallest US diocese, comprising only 179 square miles; by population, however, during McEntegart's tenure it was ranked the largest (excluding the archdioceses), with a Catholic population of over 1.5 million. Born in New York City in 1883, McEntegart was ordained a priest in 1916 for the Archdiocese of New York and won a nationwide reputation for his efforts in the field of social work. After World War II, President Roosevelt recognized McEntegart for his service in leading Catholic Charities and War Relief Services during the war years. He was elevated to bishop of Ogdensburg, in upstate New York, and ten years later, in 1953, was named rector of the Catholic University of America, putting him on the fast track for his selection as leader of the important Diocese of Brooklyn. In addition to stepping up as a leader in the ecumenical movement, he promoted a closer relationship between the clergy and the laity. He engineered a multimillion-dollar construction program, which included six high schools, a medical center, and a seminary. The *New York Times* observed in his obituary that McEntegart was "inevitably overshadowed by Cardinal Spellman, his colleague across the river,"[7] a theme that would be played out during the course of the fair.

The Vatican Pavilion Committee, a working group of eight men, four representing the Archdiocese of New York and four from the Diocese of Brooklyn, was formed to supervise the building of the pavilion, the assembling and installation of its exhibits, and the administration of the entire project. Leading this group as committee chairman was Monsignor Terence J. Cooke. Cooke began his service as chairman while holding the position of personal secretary to Cardinal Spellman. A skilled fundraiser, he became "a cash register whose keys the Cardinal pushed."[8] An excellent administrator, Cooke did not yet know that he would also direct the planning and preparation of the first papal visit to New York. By the end of the first fair season, he had been promoted to auxiliary bishop of the Archdiocese of New York, and shortly after the death of Spellman he would be named his successor as archbishop in 1968 and one year later would be elevated to cardinal. He is remembered for his tireless work with the poor and with AIDS victims and his "double life" fight against abortion as well as the death penalty.

Those assigned to the nitty-gritty responsibilities of project management had already proved their mettle administering their diocesan duties. In charge of finance was Monsignor James W. Asip, also an excellent fundraiser, who as director of the propagation of the faith raised more money per capita than any other diocesan director in the country. Monsignor Francis M. Costello, who headed the influential Archdiocesan Building Commission, was well qualified to oversee the construction of the pavilion; no archdiocesan-owned institutional construction or renovation project advanced without the approval of Costello's office. When recruited for the Pavilion Committee, Monsignor Timothy J. Flynn was already the director of the Bureau of Information and the Office of Radio and Television for the archdiocese. He also served as administrator of the United Nations Parish of the Holy Family, the church that would later host a papal visit before the fair closed.

Responsibility for the day-to-day administration of the pavilion was assigned to the well-loved Brooklyn priest, Monsignor John J. Gorman. As program director, Gorman was thrust into the limelight meeting high-profile politicians, royalty, and Hollywood movie stars, as well as administering less glamorous functions, such as lectures, Communion breakfasts, guided tours, pilgrimages, and visits of convention groups. He was assisted in the operation of the pavilion by Monsignor Joseph T. Lahey, as associate director, while Raymond S. Leonard, a priest from Brooklyn, led the preparation of the pavilion exhibits with a staff of qualified artists, designers, and technicians. To devote himself full time to this project Leonard took a temporary leave from his teaching post in the philosophy department at St. Joseph's College.[9] Each committee member was, of course, critical for the success of the pavilion, but the heaviest weight fell on the member who would be responsible for the safe transport of the most beloved fifteenth-century Renaissance masterpiece. That essential role fell to Edward M. Kinney.

As chairman of the newly formed Vatican Pavilion Transport Committee, Kinney was responsible for the shipping of the *Pietà* from St. Peter's Basilica in Vatican City to Queens, New York, and back. At the time Kinney was the general manager of the Central Purchasing Bureau of the Archdiocese of New York and the director of purchasing and shipping for Catholic Relief Services, the overseas relief and development agency of the Catholic Church in the United States. He was also tasked with running the pavilion gift shop, which unexpectedly became one of the fair's money machines.

Well before sunrise on November 1, 1962, at 3:00 a.m., Robert Moses, as president of the fair, stood with Rev. John J. Maguire, the auxiliary bishop of New York, and the pavilion directors for the official groundbreaking ceremonies at the future site of the Vatican pavilion in Flushing Meadows. In Rome, it was 9:00 a.m. when Cardinal Spellman, Bishop McEntegart of Brooklyn, and Thomas J. Deegan Jr., the chairman of the fair's executive committee, joined Pope John XXIII and Ameleto Cardinal Cicognani, papal envoy to the United States, to connect via intercontinental radio to the construction site in Queens. The event, covered by the *New York Times* on both sides of the Atlantic, coincided with the end of the first session of the Ecumenical Council. After Deegan presented the pope with a gold commemorative medal, the pope read a prepared statement (in Latin) stating that he was "spiritually present" at the groundbreaking and believed that the World's Fair would contribute "to the solidarity of peoples, and to their fruitful collaboration for the welfare of humanity."[10] With his call Pope John XXIII set in motion the start of the pile driver waiting on the other side of the Atlantic, officially opening the construction of the pavilion.

To mark the occasion, the Fair Corporation would publish a commemorative booklet, highlighting the cooperation among all involved. Behind the scenes, however, tensions were growing between the high-stature team of Cardinal Spellman, who occupied center stage whether they meant to or not, and the lesser-known but equally dedicated clergy of Brooklyn. The Brooklyn Diocese's Asip and Leonard watched the media gravitate toward the Manhattan team while their own bishop, McEntegart, was overlooked. The growing rift ruptured as the Fair Corporation filmed the groundbreaking ceremony and, for fear of being excluded, the Brooklyn team had to elbow their way up to the dais. Asip later took up the matter with Deegan, who promised more balanced coverage in the future. Asip also took the opportunity to acknowledge that Cardinal Spellman had always been attentive to include Bishop McEntegart in any function related to the fair, and that he supposed the fair personnel would do the same.[11] Such was the star power of Spellman in the early sixties.

The committee worked closely with the architects Kiff, Colean, Voss and Souder; Hurley and Hughes; and Luders Associates, and with the Stewart M. Muller Construction Company. Not surprisingly for New York City, as soon as construction started, it had to stop, owing to a three-day cement and stucco workers' strike. But work then recommenced and progressed at a good pace.

The building site was filled-in marshland, which would have to support the steel-framed pavilion building. This required a foundation consisting of 510 concrete-capped piles, each seventy feet in length. Steel-frame construction was typical of the fairground buildings. It was easy and fast to build, with little lost construction time because of the weather (there was also the scrap salvage value of the steel once the fair was over). The oval, spiral-shape design of the pavilion presented a challenge, as there were very few right angles; each steel member required meticulous design and fabrication. The floor was a concrete slab.[12] Since the construction proceeded smoothly, the pavilion directors turned their attention to the interior.

The committee was drawn to the designer and architect Charles Eames to fashion the exhibition space. Eames was the 1960s boy wonder of the design world for his groundbreaking contributions to furniture design, industrial design, and manufacturing. He and his collaborator wife, Ray, had established themselves as pavilion designers with their leading role on the US Science Pavilion at the 1962 Seattle World's Fair and had been commissioned by IBM to prepare its New York pavilion exhibition, graphics, signage, and even films.[13]

Timothy Flynn, representing the archdiocese, Asip and Leonard from Brooklyn, and Jurgen Edward "Ed" Luders, a member of the architectural team, met with Charles Eames in New York in early February 1963 to discuss Eames's plans and ideas. Right from the start, Luders, an émigré from Germany, was not impressed. Beyond requesting a plot plan for the pavilion, Eames seemed to lack interest. But two days later, when the team met for lunch at the New York Athletic Club, Eames was more engaged, and when pressed, he gave a definitive verbal agreement.

On March 13, 1963, three weeks after their first meeting, Eames presented his proposal to the directors at the Hotel Madison in Manhattan. It was a radically different design concept from the one the directors had envisioned. Instead of having a mezzanine-level chapel, Eames would bring the chapel to the ground level, flipping the original layout, with display galleries on the upper levels. Eames recommended placing figures of the twelve apostles along the roof, silhouetted against the sky and evoking St. Peter's Basilica in Rome, and a stronger "lantern dome," which would entail raising the height of the pavilion, which had already pushed the height limit established by the Fair Corporation. He proposed special liturgical music to be commissioned by the Vatican and written by a well-known composer such as Leonard Bernstein.[14] While Eames spoke, the committee members were silently calculating the additional expenses—music alone would cost a small fortune.

As they reviewed Eames's proposal, the directors knew they had a problem. On a spiritual level, the Holy Mass, the central act of worship, was too sacred to be exposed to a parade of tourists. On a practical level, Eames's design could only accommodate foot traffic of less than a thousand people per hour, while the projected number was three thousand. The directors also balked for financial reasons, as $150,000 had already been spent on architectural plans, and there had already been two revisions to the lantern dome.[15]

As the pavilion design was a joint project between the Archdiocese of New York and the Brooklyn Diocese, the opinions of directors from both dioceses had to be weighed. Once again, the issue exposed a territorial fissure—the Brooklyn directors were against the Eames plan in toto, while the New York directors were in favor of the plan in toto. With a deadlock, the final decision would be up to Cardinal Spellman—who, inconveniently, was in Rome. Monsignor Flynn caught the first flight out to review the plans to with him.[16]

Spellman was not in favor of Eames's proposal—in short, it was "not acceptable."[17] Now the directors had to explain to Eames why and determine the extent of their financial commitment for the work already completed.[18] On May 13, they met with Eames and discussed the issues, asking if he would continue on as a consultant, which he agreed to do.

Eames recommended another high-profile designer, Alexander Girard, but "warned again and again" that time was short to prepare the exhibit. Father Leonard and Ed Luders met with Girard on May 19, who advised that his fee would be between $50,000 and $100,000 and that he would like to work from his Santa Fe studio. Leonard and Luders also met with other design offices in New York. Rene D'Auriac Associates showed enthusiasm for the project, noting that their exhibit would "stress the social progress and radical movements for the benefit of mankind inaugurated by the Catholic Church." His asking price was $300,000 to $500,000.[19] The committee came to the conclusion that designers were "very expensive workmen" and that the committee would proceed using consultants and not commission any one artist or design expert. The proposals they had received were fine as far as they went but were lacking in a spirit steeped in Catholicism.

Leonard laid out two principles for the exhibit design: first, each section of the exhibit must be a unit and kept tight as a unit. Yet the viewer should get a unity of impression in viewing the whole. The viewer had to understand what was being said and understand quickly. Second, all writing that had anything to do with the pavilion had to be "controlled and of high caliber." It was increasingly felt that someone should be devoting his full time to "live with it." If so much was being spent on the building, the energy of at least one person to work completely on the exhibitions was required.[20] Furthermore, Leonard and Luders both wanted the exhibit to be beautiful, spiritual, and informative. In one last meeting before the final submission of their own plan to Spellman, Monsignor Asip suggested to Leonard and Luders that they refine and condense their notes, then Monsignor Cooke suggested they refine them again, warning the two that once the plan was shown to Spellman before he returned to Rome again, "they will be down a one-way street and there is no turning back. Therefore, we must be sure it is the right street."[21]

Once the exhibits were agreed upon, production began in a workshop in a suburb of Rome. Small-scale models of the exhibits were used for countless reviews and revisions, and once the models were finalized, life-size mock-ups were built in a barn in the town of Brewster, in rural Putnam County, New York. Here design concepts, crowd flow, and space use could be pre-tested. They could also experiment with ten simultaneous film projections against a single wall. When the mock-ups outgrew the barn, the design team moved them outdoors, creating so much interest that the team invited the local press and community for a preview.

As the pavilion construction and exhibition design progressed, the hard work of financing the project remained. At the bishops' 1962 annual meeting, Cardinal Cushing of the Archdiocese of Boston recommended that a collection for the fair be taken up in every diocese. For his part, he pledged $150,000. Cardinal Spellman wrote to every bishop in the country, issuing an appeal for contributions to defray the projected $3 million cost. Spellman echoed John XXIII's hope that "the remarkable progress of science will serve for the spiritual progress of mankind." The bishops were not happy. It wasn't easy convincing parishioners out in Iowa or Kansas that funds were needed to help out New York. Initially, Cardinal Meyer of Chicago and Cardinal McIntyre of Los Angeles did not support participation, but both eventually reversed their objections.[22]

Separately, in a pastoral letter read at Masses in all 401 New York parishes, Spellman announced that funds would be sought in every Roman Catholic diocese in the nation, and a special collection would be taken up the following Sunday, on May 26.[23] Through these efforts, Spellman raised $2.6 million, but he was still short $1.2 million. Chicago's Meyer had delayed the collection to take advantage of media coverage, and the press in Los Angeles was slow to promote the fair.[24] If they could not close the deficit, it would fall on Spellman and Bishop McEntegart of Brooklyn.

The pavilion committee could draw from a number of influential business leaders and agreed that the pavilion's official travel agent would be William Fugazy, who along with his brother, Louis, was the head of Fugazy Travel Bureau, the most ubiquitous travel agency in New York in the latter half of the twentieth century. In the metropolitan area, the name Fugazy was synonymous with travel. Other travel agents were interested, but Fugazy got the nod. With an eye on trade magazines and publicity, Fugazy asked that his picture be taken with Spellman and McEntegart, though one member objected on the grounds that Fugazy was associated with boxing promotions at a time when the sport of professional boxing had come under scrutiny for its underworld ties. Fugazy, once a boxer himself, was in favor of tighter controls. The committee overrode the objection, and the photo was approved.[25]

As the fair opening approached, the exhibitions began to take shape. Master artisans of the Vatican Museum constructed a replica of Saint

Peter's shrine-tomb to be displayed at the pavilion. It was shipped on the Italian liner *Vulcania*, leaving Naples the last week of January 1964 and arriving early February, to be transported directly to the Vatican pavilion for installation. When Pope Paul VI announced at the closing of the Second Vatican Council that he would travel from Rome to the Holy Land in early January—the first pontiff to do so—Robert Moses saw this as a good sign for the fair. Followed by television camera crews throughout Israel and Jordon, Pope Paul VI would draw attention to the early roots of Christianity, which could only increase the attraction of the shrine-tomb at the fair.[26]

The most important shipment, however, would not be so routine. Spellman assigned overall responsibility for the safe shipment of the *Pietà* from the Vatican to New York to Edward M. Kinney, the Vatican purchasing agent in the United States. Kinney was recruited to serve as the chairman of the newly formed Vatican Pavilion Transport Committee, and he came to this role with a strong résumé.

Born in New York City to Irish immigrant parents in 1913, Kinney grew up in the Bronx and graduated from the City College of New York in 1936. He studied at the School of Social Work at the Catholic University of America in Washington, DC, from 1936 to 1938, then started as a social worker but moved into administration. During World War II he joined the Catholic War Relief Services and was placed in charge of the acquisition and shipping program that sent millions of tons of food and clothing all over the world. From 1947 on, Kinney organized and supervised the primary funding of the Catholic Relief Services' multi-billion-dollar relief and development programs through the parish-wide annual Laetare collection. As assistant to the executive director of Catholic Relief Services, he built up enormous resources around the world to dispatch on short notice. His job was to prepare for disaster: earthquake, typhoon, war, flood, volcanic eruption, or tidal wave, wherever it hit and whatever it was, Kinney was ready. He would not know when it was coming, but it was coming, and there would be homeless, hungry, and helpless people, with the threat of an epidemic. Through a supply pipeline and network around the world, Kinney was able to deploy what was needed, wherever it was needed. He could move medical aid, food and shelter, jeeps, and communication equipment on the ground faster and with more accurate information than what the governments or media could provide. Additionally, he was skilled in negotiating with unions—a very valuable asset.[27]

Kinney recruited two other members for his transport committee—the first was Joseph G. Kearns, president of the D. F. Young Company, an export broker and international freight forwarder. Kearns began his career at the firm as an office boy and worked his way up to become president of the company. Kinney and Kearns met through their work for Catholic Relief Services in postwar France. Impressed by Kearns's efficient and economical work, Kinney steered thousands of CRS shipments through him, everything from "bulls to the papal farm at Castel Gandolfo, goats to Taiwan, antibiotics to Morocco for earthquake victims, blacksmith tools to Kenya for Polish refugees there, temporary emergency housing to Chile, milk to refugee children in Korea and atomic isotopes to Italy for brain tumor treatment."[28] Kearns requested the expert advice of two of his colleagues at the States Marine–Isthmian Line: senior marine architect Edgar P. Bainbridge, and Captain Philip A. Shanley, a master mariner, who through his years of experience at sea was an expert in cargo handling on the Atlantic and the Mediterranean. Kinney welcomed their assistance.

Kinney's other recruit was John Murray, vice president of the trucking firm McNally Brothers. Murray had served with the US Army as a first sergeant in peacetime Germany. Upon graduating from Fordham University in 1957 he found employment with the McNally Brothers, a trucking firm owned by his classmate's family, and, like Kearns, worked his way up to company president. Kinney wanted a committee of three, as well as a mover and a packer willing to donate their services, all of which he now had.[29]

With the committee in place, decisions could now be made. Working on a rumor, the *Washington Post* reported that the *Pietà* would make its journey across the Atlantic underwater via submarine, as it would be the "safest, smoothest and most shock-free trip."[30] The paper continued that the US Navy had accepted the mission but that it was a matter of national pride that an Italian vessel be deployed, although there were few submarines in the Italian Navy at the time.[31] In fact, Kinney's committee had already decided that their first preference was to transport both the *Pietà* and the smaller companion piece, the *Good Shepherd*, by ship rather than plane. Transatlantic flights were safe enough, but the committee knew that if the plane were to go down, the chances of survival of the statues were remote. But the possibility of submarine transport was rejected.

With the mode of transport decided, the team could turn their attention to packaging. The outer case had to be secure and made of steel, so they focused on the more difficult questions of how to package the inner crates. As luck would have it, the team learned there was a full-size plaster cast of the *Pietà* at the Metropolitan Museum of Art's storage facility in upper Manhattan, located under an elevated section of the West Side Drive. What was a replica of the *Pietà* doing in storage?

During the second half of the nineteenth century, when travel to see European sculptures was a privilege reserved only for the wealthy, acquiring plaster casts of originals was considered the next best thing. Museums and educational institutions considered plaster casts of Western European sculpture to be an essential and affordable component to building their collections.[32] American museums and collectors either bought ready-made casts of original works directly from European museums, such as the Louvre and the British Museum, or they commissioned teams of craftsmen to make the casts themselves.

At the Metropolitan Museum of Art, the plaster casts had long been on permanent exhibition in what is now the Medieval Hall, and in two adjacent wings. By the early twentieth century, interest in the cast collection began to subside as the museum could afford to buy original masterpieces and no longer had to rely on replicas.[33] Still, the collection remained on display for the enjoyment of visitors, ranging from groups of schoolchildren to the general public, until the late 1940s, when modern art turned the Western classical tradition on its ears and the casts had to go. While some were destroyed, most of the casts ended up in the museum's 158th Street storehouse. Without adequate space, the fragile sculptures were piled on top of each other and left unprotected from leaky roofs, wind, and vibrations from traffic.[34]

"After clambering over assorted statuary in various stages of repair and disrepair," Kinney recounted, "we noted with satisfaction when we reached the *Pietà*, the pyramidal mass the statue represented."[35] The men were encouraged by the wide base of sixty-three inches versus a height of sixty-nine inches, which meant a stable weight distribution. But they noticed other problem areas—the left arm and hand of the Virgin, particularly the fingers, which had been damaged from either an earlier move or plaster falling from the ceiling, then repaired or replaced by a hand different from Michelangelo's original. The extended hand would be a problem, but special care in packaging also had to be given to other vulnerable areas such as Christ's right arm, both his hands and legs, and the heads of both figures. With less than one year to go, the men had their work cut out for them. Kearns focused on the issues of marine transport, while Murray studied potential packaging methods and issues, as well as transporting the cargo on the land legs of the journey both in New York and Italy.[36]

Convinced that the solution to packaging the *Pietà* for safe transport was cushioning, Kinney and his team began investigating polystyrene foam (Styrofoam), which could mold itself to the form of the specific object. Because the committee could not test the packaging by "drop tests" or dispatching the *Pietà* on a series of hazardous trips, they did the next best thing. They visited one producer's New Jersey research center to conduct tests on a five-foot plaster statue of the *Sacred Heart*, the closest comparable object available. They supplied a wooden crate that allowed for a five-inch clearance on all sides. Once the statue was standing in the crate, covered with a vacuum-sealed polyethylene bag for protection, the liquid foam (a mixture of two basic chemicals) was poured into the wooden box.

The committee then took the crate to McNally Brothers' warehouse in Brooklyn and instructed McNally's men to subject it to the rough and tumble handling it might receive aboard a ship. This included pushing the crate off the tailgate of a truck going more than sixty miles an hour, and even dropping it off the roof of the warehouse to the street below. While the wooden crate began to splinter, it held its shape. The question was whether the statue would. After shaving off the dried foam with a two-handled wire, they could find no damage. However, perhaps owing to a faulty seal, the polyethylene bag had not prevented the foam from adhering to the statue itself, and removing the foam from the statue's surface would present other problems. The committee had to keep looking for a better solution, but they knew that they were on the right track.[37]

Around this time, James B. Gordon, a foam packaging specialist at the firm Sinclair-Koppers, came to the attention of the team. In the early 1950s, the Koppers chemical company purchased a license to manufacture polyethylene at its Port Arthur, Texas, plant. These operations later were the basis for forming a new corporate entity with Sinclair Oil Corporation, the Sinclair-Koppers Company. Initially, Gordon did not want any involvement with this project. The new packaging material, Dylite, was "pre-expanded" foam in the form of panels or beads and was used to package sensitive electronic equipment; the thought of using it to pack a three-and-a-quarter-ton marble sculpture was absurd. But the more he thought about it, the more attractive the challenge became. Gordon decided to experiment. He sent a foam-protected egg in a large box, and it arrived without a crack. He continued his egg experiments with drop tests, which also proved successful. As a packing engineer, Gordon understood that when packaging mass-produced goods, a percentage loss of replaceable goods was accepted and possibly expected, but this was not the case with works of art.[38]

The *Pietà*, moreover, was not egg-shaped. Gordon's packaging expertise, however, helped him in identifying the problems presented by the inner protection around the statue. These included the unsupported extensions of the Virgin's left hand and Christ's left foot, voids in the cape of the Virgin, and, on both figures, the relatively sharp edges usually surrounding these voids.[39] Discussion followed about leaving large void spaces around both the Virgin's hand and Christ's foot, but any whipping action, such as severe rolling of a boat, could snap off either extension from its base. The extensions would need secure bracing, and the voids would have to be filled to avoid breakage.[40]

Gordon considered using Dylite foam beads to pack the inner box, since in theory the beads would provide uniform pressure over the entire surface of the statue and would ensure that the voids would be filled. The beads would also give support to the extensions without imposing uneven force on any area. The use of the foam beads looked like an ideal solution, but the theory needed verification, so Gordon used a plaster replica of Michelangelo's *Moses*. Although the replica was smaller than the *Pietà*, there were similarities between the two works: their large, bulky bases, similar centers of gravity, and extensions that could be easily damaged. Gordon lined a plywood box with polystyrene-foam board, and the void around the figure was filled with the beads. The box was overfilled to ensure pressure on the statue; forcing the lid into place compressed the loose fill. Then followed a series of severe drops. Gordon had to investigate every possible source of damage. After each drop, he opened the exterior wooden box, and each time there was no evidence of any movement through the fill, and no damage to the statue.[41]

Dylite foam proved to be an ideal packing material for works of art. As a non-abrasive, it could be used against any surface without harm. The beads carried a negative charge, so they had a tendency to disperse, which meant that they would fill every void in the statue. They were also virtually fireproof, which would be safer than the traditional wood shavings or pressed paper. Dylite was light and buoyant enough to float; but most important, one pound per cubic foot of Dylite could absorb as much energy as eight pounds of wood shavings (excelsior). This meant its cushioning power was 800 percent greater than the favored methods of the Europeans.[42] Now Kinney just had to convince the Italians.

6

Moving Marble

Sacred Cargo

With the fair construction well underway, Moses assigned John S. Young as the liaison officer between the Fair Corporation and the papal chancery on all matters related to the Vatican pavilion. Young was a fellow Yale alumnus, and not long after graduation began working as a staff announcer for the National Broadcasting Corporation (NBC). Moses hired Young as his director of radio and television during the 1939 New York World's Fair and now again in 1964. Young (and the press) kept Moses informed of the progress in shipping the *Pietà*. For Spellman's part, his team had done all they could do to prepare stateside for the transatlantic voyage. Now it was the Italians' turn. Fortunately, when it came to moving marble, no other city had a longer history than Rome.

After the emperor Trajan defeated the Dacians in 101 and 106 AD, he commissioned a triumphal column to commemorate the victory. Completed in 113, this column stands 126 feet high and consists of twenty colossal marble drums, each weighing thirty-two tons, stacked one on top of the other. To lift the marble blocks, a system of pulleys, ropes, and capstans were employed. Centuries later, the same basic technology was used to transport and hoist ancient Egyptian and Roman obelisks, which along with the column remain standing today.

On a smaller scale, sculptors, too, had to ensure the safe transport of their marble, but without access to engineers or today's high-tech scales and lasers. Today, museums have access to the latest technology to determine the weight of immovable Renaissance sculptures. Laser scanning, yielding precise cubic dimensions, multiplied by the weight of marble (170 pounds per cubic foot), can determine a figure's weight. Michelangelo, however, had to move his *Pietà* from his workshop to the chapel at St. Peter's without such information. He probably managed this with a combination of wooden rollers, iron bars, and an "old four-wheeled iron cart."[1] In his study of Michelangelo's *David*, scholar William E. Wallace explains that sixteenth-century sculptors estimated the weight of a block of marble in *carrate*—or cartloads. In other words, the number of oxen required to pull a block on level ground was a measure of its weight.[2]

Toward the end of the nineteenth century, with the growth of art collections and the mushrooming of museums, more sophisticated shipping methods were required. Trucks, trolleys, and hydraulic lifts replaced iron carts, and blanketlike pads and padded blocks secured sculpture from harmful movement in transit. With the advent of World War II, engineers made huge improvements in developing new packing materials for the shipment of war supplies. New materials were subjected to rigorous laboratory testing, exposing them to a range of hazards, such as immersion, fire, and rough handling. The resulting fiberboard containers proved more durable—and what worked so well for shipping munitions could also work for shipping art.

To address the wartime emergency that many European museums faced when evacuating their collections to places of safety, Robert G. Rosegrant wrote the seminal guide *Packing Problems and Procedure*. For shipping large objects, such as marble sculpture, he recommended double boxing. The surface of the marble first had to be protected, then the object could be braced within the box. The inner box is placed in an outer case and is cushioned by packing material, such as wood shavings, or excelsior, as the Americans would call it.[3]

When transporting sculpture, museums began securing heavy pieces to braces within the truck wall, ensuring that the weight was evenly distributed to prevent abrasion. One of the more vulnerable points of danger in large, sculptured figures has always been the neck region, where breaks often occur. Special support is needed for the head, arms, and hands. Overpadding could be just as harmful as none at all.[4]

Above all, it was critical for the movers, for their own safety and that of the sculpture, to know exactly what they were to do before they began.[5] With that goal in mind, Edward Kinney, head of the Vatican Pavilion Transport Committee, flew to Rome in October 1963 to meet with Francesco Vacchini, the director of technical services and chief engineer for the Vatican.

While Kinney sat in Vacchini's office in the sacristy of the basilica, marveling at the art treasures surrounding him, Vacchini handed him a sketch of the *Pietà* chapel and a list of basic statistics, which Kinney had earlier requested. Together the two walked over to the chapel to inspect the sculpture. Kinney immediately saw that all the fingers on Mary's left hand had at one point been broken. Some thought the damage had occurred during one of the sculpture's previous moves, or perhaps in 1736 when plaster fell from the chapel's ceiling. Four of the five fingers were repaired with the original marble, but the replacement of the thumb, so thoroughly smashed "down to the first joint," was completed with a piece of new marble. Even then, Kinney noted, the hand's little finger was so tentatively affixed that it could be moved back and forth a good half inch. This would require special attention.[6]

During the meeting, Kinney also inquired about the statue of the *Good Shepherd*, which would accompany the *Pietà* on the voyage to New York. Vacchini explained that it was stored in the Lateran Museum, several miles away across the city of Rome, still in the case in which it had been shipped back from display at the Vatican pavilion from the Brussels World's Fair in 1958. Before Kinney returned to New York, Vacchini promised to send him their plan for packaging the *Pietà*, and two weeks later, Kinney received an outline of that plan. The Italians called for using three concentric wooden cases, with the marble sculpture braced in the innermost case. Like a Russian *matryoshka* doll, this case would be nested in another case two inches larger. Likewise, that case was to be packed in a larger third box. The spaces between the cases would be filled with wood shavings, and paper padding would be added for extra protection. This was the method art packers had used in packing statues for the last two hundred years.

In January 1964, with only four months before the fair's opening day, the transport committee made its first joint trip to Italy to make further preparations. Along with a closer examination of the *Pietà*, further discussions at the Vatican, and meetings with Vatican City and Rome police, the men needed to visit the ship selected for transporting the sculptures. A few American shipping lines had expressed their interest in the project, knowing the amount of publicity it would receive. The committee, however, felt that priority should be given to an Italian firm—but which one? Kinney relied on the Italian count Enrico Galeazzi to make the appropriate contacts.[7]

Count Galeazzi was a powerful behind-the-scenes figure in Rome and a close friend of Cardinal Eugenio Pacelli, the secretary of state of the Vatican, who would later become Pope Pius XII. Galeazzi served as the Vatican's official architect, which he was by profession, the same title that Michelangelo had held centuries before.[8] He would later run the Vatican Bank, while his half brother was the pope's physician. Galeazzi was the ultimate insider. He was selected in 1931 to serve as the Rome director of the Knights of Columbus, the American-founded Catholic men's fraternity and charity. The Knights were instrumental in funding construction projects in Rome, including much-needed playgrounds in response to a plea from Pope Benedict XV in 1920. As the governor of Vatican City during World War II, Galeazzi was sent to Washington as the Vatican's secret envoy in an appeal to spare Rome from further destruction when the Allied bombing of Rome began in 1943. In response, President Roosevelt promised the pope that Allied bombers would not go within a twenty-mile radius of Vatican City, and any bombing that might occur would be by the Germans.[9]

Galeazzi was also a close friend and confidant of Cardinal Spellman and could be counted on for support of this project. He recommended that the Americans hire Finmare, the holding company for all Italian government–owned shipping. It was the aptly named *Cristoforo Colombo*, a 29,200-ton, nine-year-old liner, that would carry, on its 172nd transatlantic voyage, the precious Vatican sculptures. With that decision made, the team turned to an American charter ship operator, the States

Marine–Isthmian Line, who were experts in cargo handling. With the help of the captain of the *Cristoforo Colombo*, States Marine monitored three voyages of the *Colombo* to learn as much as they could about the ship's normal vibrations and how it managed in rough seas.[10]

In his account of the preparations, Kinney notes the special considerations required for shipping cargo. It was in the best interest of those planning to transport high-value items to carefully plan an itinerary that would reduce the number of unnecessary stops between departure and arrival destinations and keep packing and unpacking to a minimum.[11] There were other considerations as well—on transoceanic passenger liners, priority is given to passenger comfort, with cargo relegated to holds fore and aft. But cargo freighters, in contrast to passenger liners such as the *Colombo*, were designed to carry their loads amidships, or toward the middle. On the third of the monitored voyages, the *Colombo* ran into severe weather, and the ship's forward section was "pounded for four consecutive days, hard enough . . . to bring one to one's knees."[12] That was all the information the team needed to select the after hold for the cargo's location. They now began securing a suitable crane that could lift this type of load.

Next, the committee test-drove the road from Vatican City to the port of Naples, to examine "every overhead wire, every street cable, every bridge clearance, every road crossing and almost every pothole in the streets of Naples over which the statues were expected to pass."[13] The road that would cover most of the distance was the autostrada, part of the Italian national highway system on which speed limits ranged up to eighty or ninety miles per hour.

Another concern was insurance. The other sculpture to be shipped with the *Pietà*, the *Good Shepherd*, had been previously exhibited abroad, transported by an American battleship for a Metropolitan Museum of Art exhibit in 1949.[14] To determine the proper amount of insurance for shipping both works, Kinney again turned to Count Galeazzi. Both men knew that the value of the statues was "inestimable," but Galeazzi said that "it would be of some solace to the Vatican authorities involved in permitting the statues to leave the Basilica if suitable coverage could be arranged." He pointed out that the shipment of the *Mona Lisa* from France to the National Gallery of Art in Washington had coverage of $2 million, and, after all, it was "but a painting."[15] Ultimately, the Fireman's Fund Insurance Company issued a record risk policy of $10 million, including the tightest security measures ever for a work of art.[16] The coverage included both the *Pietà* and the *Good Shepherd*, "pedestal to pedestal," from St. Peter's and the Lateran to the fairgrounds in Queens and back. A payout would be required in the event of the total loss of one or both statues. This would cover not only loss by sinking or explosion, but any kind of damage that would "render either statue unrecognizable or irreparable."[17] Coverages for both were subject, of course, to the usual condition report before and after each move of the statues. The insurance surveys were of critical importance—they were a minute examination of each statue for any surface flaws or previous damage. Photographs would be taken as evidence to ensure that neither the committee nor the insurance company would be held responsible for any preexisting damage. Any external visual flaws were to be duly recorded. But what about the internal condition? Were there any internal fissures not visible by the naked eye? Michelangelo claimed he had sculpted the *Pietà* from a perfect block of marble, but how could he have been sure?

There were two reasons why the team was also concerned about the *Good Shepherd*: the third-century marble sculpture, just over three feet high, had been found in the catacombs in thirty-four pieces and carefully reassembled in the eighteenth century; and second, the statue had not been unpacked from the original case in which it had been shipped back to Rome from the Brussels World's Fair. No one knew if it had sustained any damage.

It was possible to observe and report the condition of the exterior of the *Pietà*, but Kinney was also concerned with the marble itself. He wanted to know if there were internal fissures that, under severe pressure, could create a problem. How had the broken fingers of Mary been affixed when they were repaired? Kinney wanted to X-ray the sculpture—could that be done? How could they protect the luster of the marble of the *Pietà*? And what temperatures could marble withstand?

To answer at least one of these questions, Kinney turned to the photographic company Eastman Kodak. As a purchasing manager for the Archdiocese of New York, Kinney had made numerous valuable connections; George R. Struck, sales manager of Kodak's X-ray Division, was one of them. The archdiocese was an important client, annually purchasing thousands of dollars' worth of X-ray film for Catholic hospitals and institutions. When Kinney approached Struck about his desire to X-ray the *Pietà*, he explained that he wanted to learn as much as possible about the sculpture's internal structure. Kodak accepted the project and assigned their senior physicist, Dr. George M. Corney, to the job.[18]

Corney was unruffled by unconventional requests. He was called upon by the FBI to X-ray $100 bills to determine if they were counterfeit (they were) and assist the Rochester police in determining if the dice used in illegal gambling operations were weighted (they were).[19] Now he was leading a team of radiologists, equipped with a conventional roentgen ray unit and a cobalt-60 radiation source, to St. Peter's. Offering their services gratis, the team hoped to gain experience and the opportunity to take still photos, to film the radiography for their research library, and to publish their findings internally. In return, Kodak asked for a written statement of indemnity from any personal or property damage resulting from the radiography. This got Kinney to wondering what kind of damage could occur. Struck explained that "in the more than four and a half centuries of its existence [the statue] had already received from cosmic rays alone, without ill effect, more than ten times the radiation he contemplated using." To avoid any unnecessary radiation to humans, the entire operation would be held after the basilica was closed to the public, the evenings of March 23, 24, and 25.[20]

The Vatican agreed to this potentially hazardous undertaking but had two stipulations, the first of which was that there be no publicity until after the project was completed. This was to respect the sensitivity of Italians, who understandably would be concerned for the safety of the statue. Only then would the results be released. The second condition was that Kinney meet with Vatican officials to complete all preparations prior to the arrival of the Kodak unit. In the meantime, Kinney, back at his home in Chappaqua, New York, took a small replica of the *Pietà* and, with red nail polish, marked the areas he deemed most vulnerable and of greatest concern. He then sent it off to Rochester for George Corney's study.[21]

Corney and his team started their work in St. Peter's the evening of Monday, March 23—he was only able to take pictures for three hours on four days. It happened to be Holy Week, when all statues were veiled. "When

I first saw the *Pietà* in Rome," Corney says, "I froze in my tracks. The statue was completely covered with a [purple] Lenten shroud. I thought for a moment that we would be forced to make our exposures without ever seeing or examining the statue itself." When he and his crew were ready to begin their work, a temporary partition was set up, and the shroud was removed. They began their work each evening at six-thirty, as soon as the basilica was cleared of visitors, and spent two hours taking X-ray photos. Then they arranged a ten-hour overnight exposure with a cobalt isotope that emitted gamma rays. From St. Peter's, they would go directly to Kodak's office in Rome to process the X-ray photos they just took. What Corney learned from these first images was that the marble of the *Pietà* was denser than the marble he experimented with in Rochester. The next morning, they returned to St. Peter's at six o'clock to stop the gamma-ray exposure and were able to take additional pictures before the *Pietà* had to be reshrouded.

The X-rays revealed that five fingers of the left hand of Mary and the little finger of the right hand of the figure of Christ had broken off and been repaired by pins inserted into holes drilled into the marble hands and fingers. He also found two holes drilled into the Virgin's head by a well-intentioned "vandal" to hold a crown, long since removed. Corney called it "the most challenging and fascinating project [he had] undertaken in 25 years of radiographic work."[22]

On the third and final night, the team took photos of every inch of the surface of the sculpture using a Kodak Startech camera made specifically for close-up dental and medical photography. These detailed images were used for insurance purposes, as well as protection for the transport committee. The only unexpected discovery was a "curious cone-shaped void" at the bottom of the base, extending up about twelve inches, which must have been natural to the block of marble.[23] While Vacchini, the Vatican's chief engineer, was pleased with the radiography, he was becoming increasingly agitated over the American packaging proposal and filed a report to the Vatican Commission on Fine Arts condemning the process. Probably owing to the intercession of Count Galeazzi, the report was rejected.[24] Already word was leaking out of a dispute between the Italian packing team, the Montenovi family, and the Americans. On its front page, *Il Messaggero* reported that the dispute had delayed the start of packing, which in turn meant a delay of the shipment of the statue from Naples scheduled for April 5. Pietro Montenovi told the press that "everything is up in the air—I don't even know if the *Pietà* will go to New York."[25] In a rushed memorandum to Charles Poletti in New York, John Young, Moses's liaison with the chancery, detailed the problems that Spellman's team was experiencing with the movers and included the understatement, "Somewhere along the line, a slight dispute arose."[26]

To the American team it was obvious that the innovative packing methods were far superior to the traditional practice of wrapping the sculpture in linen and cushioning it with paper and wood shavings. In an interview with *L'Osservatore della Dominica*, Vacchini envisioned that the voids in the statue would be filled with gauze and that a "glove" of pressed paper would enclose the entire surface, which then would be wrapped in thin linen. The statue itself would be braced and chocked. Chocking was a method to prevent movement by placing strategic blocks or wedges under and against the statue. The Italians were not convinced modern packing methods would be better, even as the catastrophe of the *Venus de Milo* was unfolding.

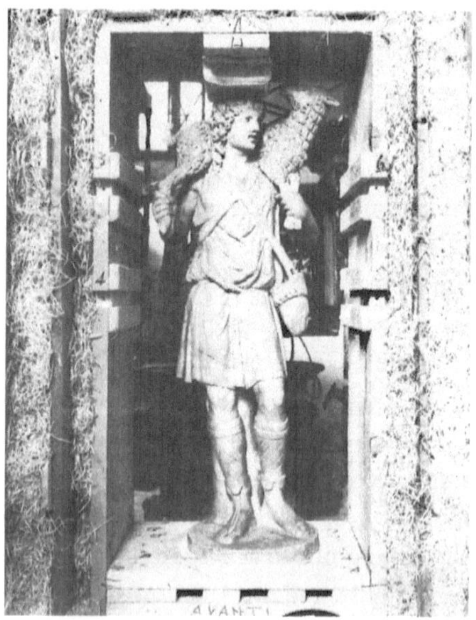

Figure 6.1. In a traditional European packing method, the *Good Shepherd* statue is braced to the inner wooden box, with additional packing boxes cushioned by wood shavings. (Edward M. Kinney, *The Saga of a Statue*)

In honor of the Olympic Games to be held in Tokyo that year, the Louvre had agreed to send its most treasured statue, the *Venus de Milo*, to Japan for exhibition in Tokyo and Kyoto. The ancient 1.5-ton Greek statue was created between 130 and 100 BC and had been discovered by a peasant farmer on the Greek island of Melos in 1820. On February 15, 1964, the *Venus* was removed from its pedestal, crated, and transported to Marseille, where it was put in the hold of the French steamship *Vietnam*. Supervising the operation during both the loading in Marseille and the unloading in Yokohama were the chief of the Louvre's marble workshop and a museum director. The two did not, however, accompany the crate on the ship. Despite anxieties that some held regarding shipping the *Venus*, one museum expert remarked of the armless statue that "everything that could be broken is already broken."[27]

After its month-long sea voyage, the statue arrived at the Museum of Western Art in Yokohama. Once the statue was unpacked, four small pieces were discovered, chipped from the folds of the robe. Though three of the pieces were plaster from an earlier restoration, one was original marble, but even that piece had been found already broken off when the statue was discovered in 1820. While the damage was minor, the tight packing of the crate, lined with soft plastic, was proof that standard packaging would not be sufficient. John Young, who happened to be in Rome when the story broke, telegrammed Poletti to report that all the Roman newspapers carried the story, and *Il Tempo* had published a front-page editorial criticizing the *Pietà* loan. Young added that this was disturbing but "does not alter our schedule at all."[28]

For generations, the Vatican relied on the packing expertise of one family, the Montenovi of Rome, who had been responsible for the packaging of precious artwork since 1870. Pietro Montenovi prepared the Vatican plan for packing the *Pietà*, which called for the sculpture to be packaged in the traditional method, cushioned with sawdust and wood shavings. Montenovi had already constructed the three boxes, with the two inner crates each two inches larger than the one inside it. The inner wooden box would be placed in a second wooden box, which would be packed in a floating steel outer box. Vatican assistants carried out a series of trial runs using a *Pietà* replica. The Montenovi family had used this method to pack art for four generations, basically unchanged from the time of Napoleon.

As the final preparations for shipment were being made, Kinney once again called the members of the Vatican Pavilion Transport Committee to Rome. This included the core members John Murray of McNally Brothers Transportation and Joseph Kearns of the Daniel F. Young shipping brokerage, plus James Gordon of Sinclair-Koppers and Edgar Bainbridge. As any experienced shipper will attest, working with a known and trusted team is paramount. "Once chosen, neither packer nor shipper can be changed without severe risks," wrote Walter Persegati in describing his later experience shipping art from the Vatican Museums.

By Thursday, March 26, the American and Italians had reached a compromise—both Dylite foam and wood shavings would be used. What this actually meant was that Kinney compromised on the packaging of the *Good Shepherd*. It would be packed in the same container in which it was returned from Brussels, but they would insert a four-inch-wide foam board between the inner case and the container. But as far as the *Pietà* went, it would be packed the "American way." This was also the approach approved by the American insurers, so there could be no compromise. To counter what Kinney considered erroneous and exaggerated press coverage by the Italians, he felt it necessary to issue a press statement clarifying the final arrangements and downplaying any differences in opinion. After attending early morning Mass at the basilica on March 31, Kinney returned to the Hotel Michelangelo for breakfast to receive a hand-delivered message from Vacchini, the Vatican's chief engineer, stating that he was withdrawing from the project. He wrote that "you have your technicians with you and by now are familiar with the various Vatican offices. Consequently, you will be able to proceed with your work, namely, the moving of the *Pietà*." Kinney and his team had anticipated practically every conceivable obstacle, except this one. Kinney acknowledged that no one even knew where the Vatican workers kept the screwdrivers, let alone the moving equipment.[29]

The next day, the *London Daily Telegraph* and others in the European press reported that Pietro Montenovi and his Vatican assistants had withdrawn from the assignment to pack the sculpture for shipment, but they had completed the preliminary work for removing the three-and-a-quarter-ton marble sculpture from its pedestal. Kinney headed over to the *Pietà* chapel, where packing was scheduled to begin at 8:00 a.m. He arrived to plenty of reporters and cameramen, but no Vacchini and no Montenovi.

Pressed by the media to explain, Kinney, although reputed to be always diplomatic, was reported to have said, "We have the responsibility, and it will be packed our way and only our way."[30] In an interview with Reuters, Kinney was "rather adamant" about the use of plastic packing, because testing had proven that it would "withstand eight and a half times more energy than wood shaving." Still, the Italians protested that old-fashioned wood shavings were still better than plastic, which they believed could damage the marble.[31]

Kinney left the chapel and headed straight for the office of Count Galeazzi, who arranged for an alternative packer, Tito Minguzzo. Minguzzo and his men could package the statue and move the sculpture to the portico of the basilica, but the inner cases still had to be built, and fast, with no time to spare.[32] After regrouping over lunch, Kinney noticed Vacchini's black Mercedes parked outside the chancery, so he proceeded directly to Vacchini's office, and contrary to *Il Tempo*'s earlier report of illness found Vacchini in a more cooperative mood. Someone had leaned on him—it could have been the Fireman's Fund, the insurer of the *Pietà*, insisting its coverage against loss or damage to the statue would depend on whether the recommendations of the American transport team were faithfully followed.[33] Or perhaps it was Galeazzi.

Whatever the reason was, Vacchini turned to Signors Pietrantonio and Ricci, the carpenters responsible for constructing seating for the Vatican Council. Kinney needed three carpenters to be at the chapel by four o'clock the following afternoon to start work, and if they could finish the job on time there would be a bonus in it for them. Avoiding a last-minute catastrophe, and when everyone had a chance to recover, Kinney graciously invited Vacchini and his wife to accompany the *Pietà* on the *Cristoforo Colombo* and to be his guests at the fair. Vacchini accepted. The following day, *Il Messaggero* reported that the dispute between the Americans and Italians had ended—the statues would be packed in Dylite foam, similar to Styrofoam but stronger and more durable, and "the wood shavings, the cork, the cardboard and the straw would be sent back to the attic." It was a backhanded compliment, acknowledging that the Italian approach to packing was now obsolete.

As soon as the insurance inspector signed off on the new materials, the packing began, with the scaffolding erected in front of the pedestal on which the *Pietà* rests. The inner-case base was positioned on this scaffolding, covered with a layer of rubber vibration pads to dampen any vertical shocks. A sheet of plywood was placed on top of the vibration pads, and on top of that two five-by-twenty-centimeter wood boards or "slipsticks" were positioned. Minguzzi's men waxed two additional boards of the same dimensions, and while one team moved the sculpture eight feet forward onto the scaffold, another team meticulously slid the slipsticks under the sculpture's base. With that, the men pushed the *Pietà* onto the boards of the inner-case base. The void created by these two boards was filled with foam board to give added support and stability to the sculpture and to prevent any loose foam particles from shifting down.[34]

Count Galeazzi had decided that the plaster addition to the original base of the *Pietà* should be removed so that sculpture could be viewed in the position that Michelangelo intended.[35] The committee had already considered the most satisfactory placement angle when chocking the *Pietà* on its slipsticks, to ensure its uneven base was supported and to minimize stress to any part of the statue. In this same manner the *Pietà* was to be

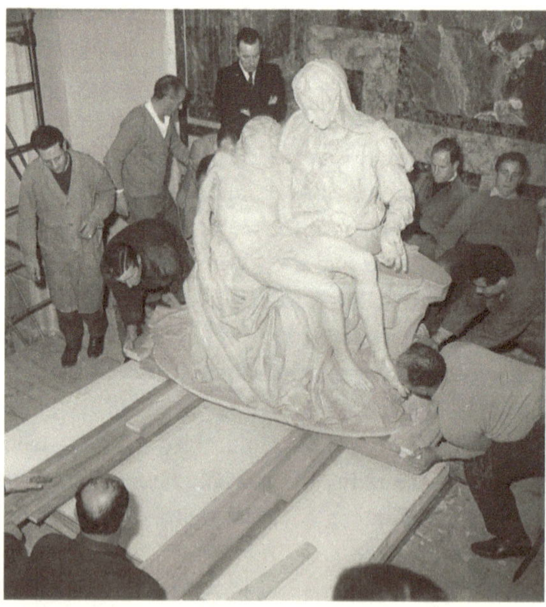

Figure 6.2. John Murray (*behind the sculpture*) supervises the packing of the *Pietà* in St. Peter's Basilica. (Photo courtesy of John Murray Jr.)

placed atop its pedestal in the Vatican pavilion. *The New York Times* reported on April 1 that "the delicate task of transferring Michelangelo's 'La Pietà' from its pedestal to the thick wooden base of its inner case was accomplished without mishap this evening." The next step would be to attach the sides of the case and complete the fill.[36]

The challenge now was supporting Mary's left hand. Two hundred years ago, when the statue was last moved, the thumb and four fingers of the hand had been broken. These breaks remained visible, and Eastman Kodak's X-ray survey had revealed the steel pins used to repair them. Carefully, as if the hand were human, the fingers were now individually wrapped in Ace bandages that Kinney had brought from the States. Strips of foam polystyrene were placed and secured between the fingers, with special care given to the little finger of the hand, which unlike the other fingers was alarmingly unstable. Once this preliminary wrapping was completed, the left hand was completely bandaged up to the wrist—looking, as Kinney observed, as though it "wore a boxing glove."[37]

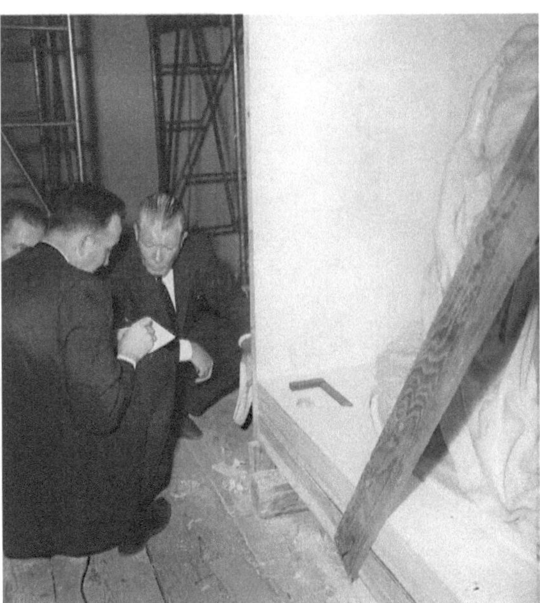

Figure 6.3. Edward M. Kinney, *left*, review notes with John Murray. (Photo courtesy of John Murray Jr.)

At this point, the news media was permitted into the chapel to observe and record the crating of the sculpture. The team constructed a wooden frame for the statue, working up from the base of the crate, careful to maintain a margin of at least 2.5 centimeters so there was no direct contact between marble and wood. With the frame in place, the two wooden side panels went up, with two boards attached to the tops of each side as a temporary brace to ensure that neither side would fall in. Then the front and the back panels were added. As an inner cushion, a foam board ten centimeters thick had been fitted against the plywood. Now the case was ready for the fill. The foam beads were poured in, section by section. Because the Dylite foam beads had a negative charge, they had a tendency to disperse, helping to ensure that every nook and cranny of the statue's surface was filled.[38]

The men kept filling the voids until finally the head of the Virgin was covered and no longer visible. Once the crate was overfilled with Dylite foam by about two inches, the final lid was put in position, and four men stood on top of it while others tightened it in place. This would create

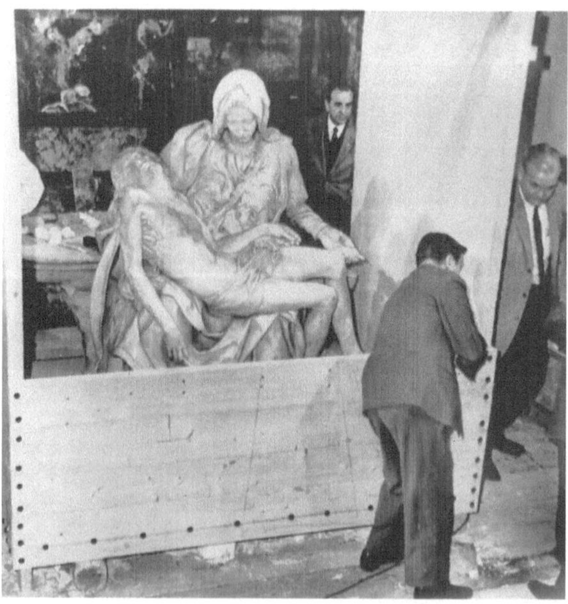

Figure 6.4. Workers assemble the wooden packing crate around the sculpture. (AP wire photo)

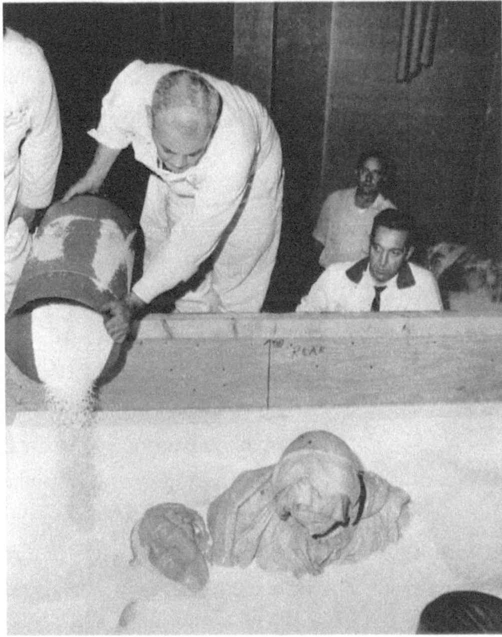

Figure 6.5. Workers fill the inner crate with Dylite foam beads. (AP wire photo)

sufficient pressure inside the crate to hold the statue firmly in place, and all that was left was to double-check every lug bolt and screw.

The inner case was now ready to move to the basilica's portico, where a crew was standing by. The crate was loaded onto a trolley-like *carrello* and wheeled up and down ramps placed over the steps and through the magnificent central bronze door, the Filarete door, which had been salvaged from the original Constantinian basilica before it was demolished in the sixteenth century to make way for the new St. Peter's. From there it was wheeled over and down the entry steps on a specially built ramp and through the guard's gate, through two of the Bernini columns, to where the steel shipping container and the crane-equipped Gondrand truck and the flatbed, "Tigre," were waiting. The steel container was lined with an asbestos sheet to protect against fire, and its bottom was covered with foam board. A plywood sheet covered the foam board to ensure uniform distribution of the statue's weight, while it also gave a smooth platform for sliding the inner wooden case onto. With the help of bars and steel rollers, the wooden crate was loaded inside the steel container, and any extra voids between the two were filled with more foam board. Together the container and the marble sculpture now weighed approximately five tons, but because a total of four hundred pounds of foam polystyrene (foam particles and board) were used, the container was so buoyant that it could float.[39]

Gondrand's mobile crane then lifted the steel container and slowly swung it over onto the waiting flatbed truck. It was already nighttime, and the journey to Naples would not start until the next morning. Now loaded, the truck drove into the Vatican City courtyard, and for the first time since the sixteenth-century day it was installed, the *Pietà* spent the night outside the basilica.[40]

As for the *Good Shepherd*, after its inspection for insurance purposes, the men had decided to reuse the same wooden crate in which it had last been packed but repack it "American style" with foam board to cushion the inner crate within its outer case, as was done with the *Pietà*. Both statues were packed without a middle crate, allowing for extra cushioning. The *Good Shepherd* was loaded separately into a smaller truck that would join the Rome–Naples convoy the next morning.

The English-language *Rome Daily American* newspaper summed up the operation on its editorial page: "The great debate over the manner in which to pack Michelangelo's masterpiece has ended. It will be transported to the United States tucked away in a wholly American plastic package—no stuffing of sawdust and wood shavings, a system used for years by the Vatican. No question now of where the blame will fall should anything go wrong."

On April 4 at 11:00 a.m., Kinney, Kearns, Murray, Shanley, Bainbridge, and Gordon gathered for an audience with Paul VI. It was Monsignor Paul Marcinkus, the powerful prelate from Cicero, Illinois, who introduced the team to the pope. Equipped with the papal blessing, the Americans were now fully ready to begin their journey.

Finally, the convoy set out, escorted by a Rome motorcycle policeman at both front and rear. Next were three Mercedes sedans, courtesy of the Vatican, then the Tigre truck carrying the *Pietà* and the smaller truck with the *Good Shepherd*, followed by an additional car for the Gondrand team. The little convoy drove the ring road around Rome, and once it entered the autostrada, the highway motorcycle police replaced the Rome police as escorts. Proceeding at twenty to twenty-five miles per hour, the team arrived in Naples after a ten-hour drive. The Tigre truck was white, with the word PIETÀ painted in blue on it, and as the convoy approached its destination, waiting Neapolitans greeted it with applause and shouts of "Viva La Pietà!" By 10 p.m. the *Pietà* was at dockside, ready for the *Cristoforo Colombo*, which would arrive the next morning.[41]

Right on schedule, on April 5, the crane lifted the *Good Shepherd* in a cargo net and lowered it in place onto the deck of the *Colombo*. Next came the *Pietà*. While the crew was checking the cables, a helicopter hovered overhead taking photographs for the press and television news. Onlookers cheered when the *Pietà* was lifted, slowly swung over the cabin-class deck, and just as slowly lowered until it was not quite touching the deck, to guide the container into the prepared deck shoes. Explosive bolts assured that if the ship were to go down the container would be released, allowing it to float to the surface. The container was also furnished with dye markers and a sonar pinger that would be activated upon contact with salt water.[42] Securing cables were affixed, and the flashing light/signal device attached and tested. Now, by 11:15 a.m., the passengers were allowed to board. Some members of the transport committee were able to fly back to the States, while Kinney, Kearns, and Murray stayed on the ship to accompany the *Pietà* the entire distance to the fair.

Also on board was Monsignor Andrew P. Landi, the assistant director of Catholic Relief Services, whom Kinney had requested to serve as the team's chaplain, offering Mass in the ship's chapel. When Kinney returned to the cabin deck to check on the container, he noticed that someone had tied a Saint Christopher medal to the handle of one of the doors of the *Pietà* case. That someone was Ed Bainbridge, the only non-Catholic on the team.[43] The *Colombo* began its westward voyage in calm waters. The first stop was Gibraltar, but with only two hours in port, passengers stayed on board and enjoyed the view from the deck. Kinney knew that over the seven-day passage the salt air and ocean spray would wreak havoc on the steel case's finish. He had already seen the results when the case was shipped from the US to Rome in March. He wanted to make sure that when they arrived in New York, to be met by a crowd of reporters and photographers, the case would be camera ready.

The biggest concern now was the weather. Two of the team members, Bainbridge and Shanley, had arranged for weather reports from a private service, States Marine, to supplement Captain Giuseppe Soletti's own reports. Based on these combined forecasts, Soletti agreed to change course slightly to avoid rough weather, veering south of the Azores by a hundred miles. Under normal circumstances the forecasts would not have been enough to warrant a change of course, but these were not normal circumstances. The remainder of the trip was smooth sailing, and each day brought them closer to warmer spring weather. As long as the weather held, the *Colombo* would arrive in New York on schedule, Monday morning at eight o'clock—and it did.

The barge-derrick the *Challenger* was in place and ready when the *Colombo* berthed. Bishop McEntegart, along with the fair's executive committee chairman Thomas J. Deegan and pavilion administrator Christopher Kiernan, who had greeted the ship on the tugboat, were now on board the *Challenger*. And just minutes after the *Pietà* had been placed on the barge, Cardinal Spellman arrived and was met by three pavilion hostesses, adding a note of glamour to the welcome. When asked if he

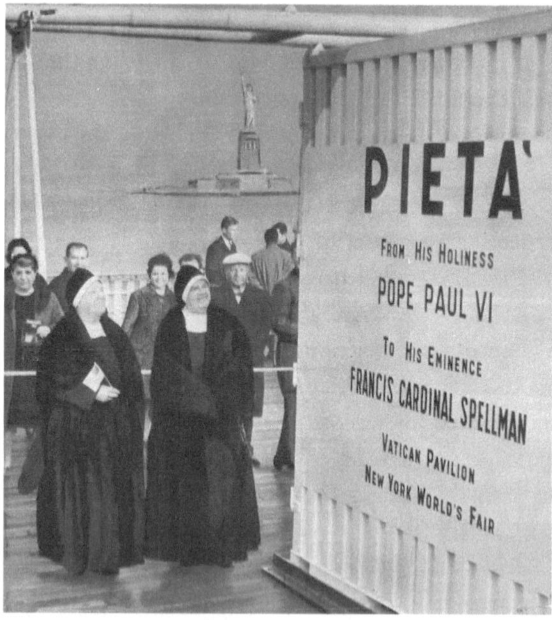

Figure 6.6. Two nuns aboard the Italian liner *Cristoforo Colombo* view the container holding the *Pietà* as the ship approaches New York Harbor. (AP wire photo)

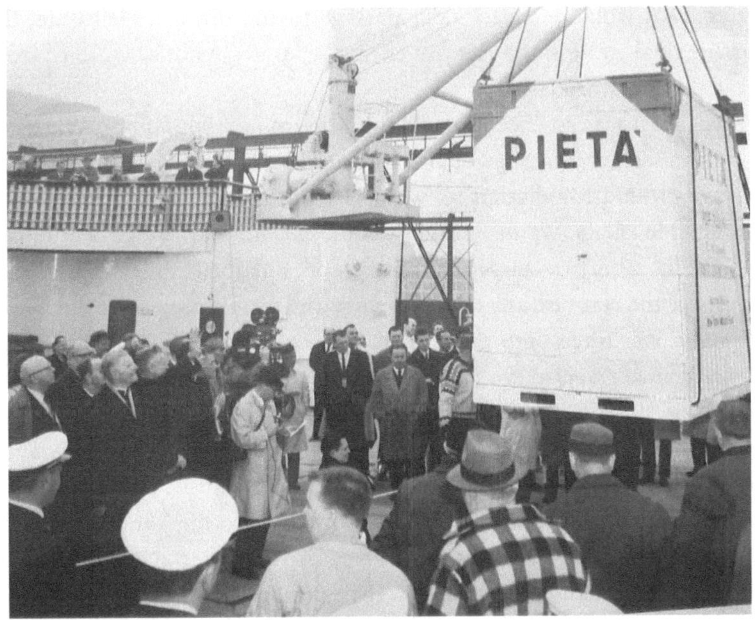

Figure 6.7. Bishop McEntegart (*left, in clerical collar*) waves to welcome the arrival of the *Pietà*. (Photo courtesy of the Diocese of Brooklyn Archives)

had come to bless the sculpture, Spellman said that "he would rather bless those who had brought it over to safety." He blessed all those present, including the two sculptures.[44]

As soon as the transfer to the *Challenger* had been completed, Captain Soletti of the *Colombo*, the *New York Times* reported, "appeared visibly relieved." The *Pietà*, inside its shipping container, weighed 11,600 pounds, just short of six tons, but the *Challenger* was able to lift ninety tons, so Ben Olsen, its captain, felt confident. The same could not be said of G. Doane McCarthy Jr., vice president of the Fireman's Fund Insurance Company. "You people can afford to be calm, but I've got $6 million hanging by those cables up there." The *Times* reporter, Robert Alden, overheard one bystander comment, "This is more priceless than the priceless *Mona Lisa*."[45]

The *Challenger* remained docked at the West 44th Street pier until the next morning, when there would be a slack tide, a roughly three-hour period between the rising and falling of each tide, with little movement of water. That would enable a safer passage to and through Hell Gate, the narrow tidal strait in the East River that separates Astoria, Queens, from Randalls and Wards Islands.

Because it was an exceptionally foggy morning, thousands of Manhattanites missed the passage of the special visitor as the *Challenger* made its way around Manhattan to Wards Island, then took its turn to pass through Hell Gate. Waiting at the Whitestone Parkway Bridge over Flushing Creek were two McNally trucks, one a flatbed trailer that would transport the statutes, the other to carry the team's gear. First the *Good Shepherd* was lifted onto the flatbed, then the "six-million-dollar baby," the nickname given to the *Pietà*, in reference to the amount of insurance issued for this trip.[46]

With trucks loaded and accompanied by a police escort, the convoy moved slowly to the fairgrounds, less than a mile away. Past the gate to the fairgrounds it was another quarter of a mile to the Vatican pavilion. They had come so far, and now all that was left was to move the statues into the pavilion and place them on their pedestals.

When the *Pietà* had been loaded onto the flatbed trailer, Murray had seen to it that the statue faced to the front of the truck, eliminating any unnecessary moves to turn it around once it arrived at the pavilion. Edmund P. Farrell, McNally's chief of operations, directed the use of rollers and a heavy chain as a forklift pulled the steel container halfway down the truck ramp; then the wooden crate was carefully pulled out of the steel case, again with rollers, just clearing the pavilion doors. From there a heavy-duty forklift lifted and carried the inner case onto a temporary platform built to surround the marble pedestal. While it sat on the platform, the inner wooden case was opened, and the team could see that the foam beads were undisturbed, looking just as they had when the crate had been packed. The crew used a commercial vacuum cleaner to remove the beads, and slowly, Mary's head emerged. On hand were two insurance inspectors, one from the Fireman's Fund to make sure no damage had occurred during transport, and another from the World's Fair Fine Arts insurance pool, which would be responsible for coverage the minute the sculpture was on its pedestal.[47]

Using the same slipsticks Minguzzi and his men had supplied to remove the statue from its pedestal in St. Peter's, the crew moved the marble sculpture from the base of the wooden box onto its new pedestal.[48] The entire team, along with the McNally crew, eased the statue to its proper position on the pedestal. Amid general joy and relief, Kinney found the nearest phone and called the cardinal's residence to relay the good news. The next morning, Kinney oversaw the unpacking of the *Good Shepherd*, which proved more difficult than unpacking the *Pietà*. Instead of vacuuming up foam beads, they had to remove the inner chocks and foam boards, to enable a clean report by the insurance inspectors.

With both statues now in place, the only thing standing between the pavilion and the crowds was the dedication and unveiling of the *Pietà*. Surely, Kinney thought, this was "one of the most carefully planned journeys in history."[49]

7

On Stage

Mielziner's Magnum Opus

In the words of the former *New York Times* theater critic Frank Rich, "Of the many fine designers who came of age with the modern theater in twentieth century, few, if any, were greater than Joseph 'Jo' Mielziner."[1] Mielziner's settings for three hundred dramas, musical comedies, ballets, and operas earned him five Tony Awards, five Donaldson Awards, and one Academy Award.[2] Now Mielziner would set the stage not for a show on Broadway but for the *Pietà* at the Vatican pavilion.

Mielziner was born of American parents in Paris in 1901. His mother was the former Ella MacKenna Friend, the Paris correspondent for *Vogue* and the first woman member of the Foreign Press Association.[3] She was a descendant of John Friend, believed to be the master builder for the original Harvard College, and of Charlotte Cushman, "brightest star of the nineteenth-century American theater." Mielziner's father, Leo, was a noted portrait painter and son of the Talmudic scholar and writer Rabbi Moses Mielziner, a president of Hebrew Union College in Cincinnati and a founder of Reformed Judaism. His great-grandfather was Rabbi Benjamin of Mielzyn, then in Russian Poland, fondly referred to as "The Mielziner."[4] His brother, Leo Jr., also in the theater world, adopted the acting name Kenneth MacKenna, using his mother's maiden name. Both brothers were referred to as *mischlinge*, or of mixed blood. In 1909, his family returned from Paris to New York, where Mielziner attended high school until he won a scholarship to study painting at the Pennsylvania Academy of Fine Arts when he was fifteen. With the outbreak of World War I two years later, Mielziner enlisted in the Marines and served briefly before returning to the academy. As his studies continued, he realized painting was not his calling, but that stage design was. He won two scholarships to travel in Europe, where he took in the tremendous changes occurring in traditional stage design, visiting thirteen countries in thirteen months. He absorbed the leading-edge work of composer Kurt Weill, playwright Bertolt Brecht, and director Fritz Lange, and while working for the Austrian architect and set designer Kurt Strnad at the Kunstgewerbeschule in Vienna, Mielziner absorbed the theories of the modernist New Stagecraft. Settings became "silent characters." From this Mielziner developed his own approach, which he described as "implied scenery," simple settings that were suggestive, not literal.[5]

Equipped with fresh ideas, Mielziner began his stage design career in 1921, and within two decades he was well established in the theater world. Mielziner's biographer, the theater scholar Mary C. Henderson, wrote that he "mastered the technique of 'cinematic' scenery using the tools that everyone had at hand but using them better."[6]

Though he acknowledged his mentors, Mielziner considered himself a self-taught lighting expert. He commented that "modern lighting effects are wonderful—so subtle, changeable, exciting," and especially valued the dimmer.[7] The development of the dimmer, used to vary the intensity of the light, was revolutionary in stage design. Mielziner always insisted on designing the lighting for all his shows rather than relying on a specialized lighting designer.[8] Known for his ingenious designs, Mielziner could command higher fees, and producers knew his name would lend their production a special cachet. He was not only skilled but also reliable, having developed a reputation for being always prepared and always on time. These qualities made him an attractive partner.[9] He designed not only for the stage, but his skills were also in demand as television audiences grew; he was hired as a consultant for CBS-TV.[10]

His success, however, took a heavy toll on his personal life, and his search for a "philosophical center" led Mielziner to the Catholic Church.[11] A friend who was undergoing instruction in the Catholic faith invited Mielziner to Mass at the Church of St. Paul the Apostle on the corner of Columbus Avenue and 60th Street. Spiritually hungry, he listened with an open mind and heart to the message given by the priest from the pulpit, Father Fulton Sheen. After six months of instruction, there remained only one obstacle to his reception into the Catholic Church, and that was the matter of his divorce. Father Sheen explained, however, that because neither Mielziner nor his wife was baptized at the time of their marriage, he would be eligible for an allowance by the church for the dissolution of the marriage, referred to as the Pauline Privilege. Still somewhat resistant, he balked at what he viewed as a limitation on his intellectual freedom. Sheen did not argue, he merely asked, "And are there no limitations which you find it necessary to observe in your own work?" Of course there were.

Mielziner had already read and been influenced by G. K. Chesterton's book *Orthodoxy*. Chesterton had written of his own conversion, and he called on others to live in the world and not to retreat from it. Mielziner commented in one of his letters that

> [it was] a period in my life when I was going through an internal struggle which, in a sense, was the desire to embrace a strong faith that would make demands on my life. In opposition to this was my concern and fear and doubts that accepting the dogma of, in my case, Catholicism would restrict what I felt was freedom of thought, and the dangers of a closed mind on the social and intellectual level. In Orthodoxy, Chesterton showed me the answer. In this book he convinced me, to my complete satisfaction, that in accepting a creed, I gained a freedom rather than lost one. It was a freedom I did not enjoy previous to the time of accepting a definite faith.[12]

Satisfied that conversion would enhance his intellect, Mielziner was received into the church on Christmas Day of 1936 by Father Sheen in New York's Church of the Blessed Sacrament. He was thirty-five years old. Of his conversion, his father said to him, "Since you cannot stay with the religion of your ancestors, I am glad you're going into a real one. I have only one request. As long as you live, take pride in your Jewish heritage. There are many things in it we can be proud of."[13]

Part of Catholicism's attraction for Mielziner was its discipline. In an interview, Mielziner reflected on the value of religion for those with creative gifts: "There is so much ego in the theater, so much temperament—people tend to forget that the talent they have is a gift, on lend-lease from a higher authority. It's important if they can be reminded through religion."[14] Sheen remained Mielziner's spiritual adviser and close friend throughout his life.[15] Later named an archbishop, Sheen may be best remembered for his broadcast *Life Is Worth Living*, the popular inspirational television series that was watched by millions of viewers from all walks of life, regardless of religious belief, from 1952 until 1957.

With the advent of World War II, Mielziner once again enlisted, this time with the Army Air Forces, and, with his expertise, offered to develop camouflage, earning him the rank of major. He commented that "on the stage my job was to make people grasp a situation as quickly as possible. In camouflage, my job was to keep them from grasping it at all."[16] But over time his enthusiasm dampened, and with the assistance of Sheen and other well-placed contacts, he was transferred to Washington's *Office of Strategic Services*(*OSS*), a wartime intelligence agency.

While he was stationed at Jefferson Barracks near Saint Louis, Mielziner met Father Stanley Kusman, a Marianist priest.[17] Kusman, whose charisma and dedication earned him the sobriquet "the poor man's Bishop Sheen," enlisted as an army chaplain and was assigned to the Italian campaign.[18] Mielziner seemed to have an uncanny ability to bond with the most exceptional priests, and he and Kusman developed a friendship that would last till Kusman's death.[19] Mielziner's deep spirituality joined with his creative talent could have potentially found its fullest expression in the church rather than the theater. In a 1955 interview, Mielziner commented, "If I were starting from scratch, I'd go to a Catholic school, the best of its kind I could find. I'd study church architecture, church structure, and décor. Then I'd devote my working efforts to the liturgical arts. As a matter of fact, I may take a whack at liturgical art yet. Not tomorrow, mind you. Say ten or twelve years from now, after the kids have grown up and I no longer have to keep my nose to the grindstone."[20] With the announcement that the Vatican pavilion would host the *Pietà*, Mielziner now had the chance "to take a whack." As it happened, Father Kusman was leading a retreat in Rome in 1960 when he learned that Cardinal Spellman had requested the *Pietà* loan from Pope John XXIII. As soon as he returned to the States, he said to Mielziner, "Why not ask the Cardinal if you can submit a design?" Within an hour, Cardinal Spellman had phoned his reply—he enthusiastically accepted. Little did Mielziner realize that his offer would consume almost a year of his time. He worked on preliminary sketches until 2:00 a.m. that night. After a number of sketches, he arrived at his final design—the one critics hated and the people loved.[21]

Mielziner offered his services as a designer and lighting expert gratis to prepare an appropriate setting worthy of the masterpiece. At the end of his formal acceptance letter to Spellman, he added that they had met once before, in 1947 when Mielziner had received an honorary doctorate of fine arts at Fordham University, where Spellman was presiding over the graduation ceremonies.[22] Mielziner considered those in the theater to be particularly equipped for the task he faced, but stated that "the exhibit is being treated like an altar—so you don't feel as if you're going into another museum."[23] He did not want to present the sculpture out of context.

Mielziner had two objectives in designing the setting for the *Pietà*: he wanted the angle of observation by the viewers to be as close as possible to the sculptor's original intention, and he wanted to meet Cardinal Spellman's approval.[24] Mielziner understood that Michelangelo sculpted this masterpiece as a work of beauty that was to serve a liturgical purpose as an altarpiece. He could have designed an appropriate Baroque background, but how could one match the magnificence of St. Peter's interior? Mielziner never considered replicating the St. Peter's chapel setting. He chose an entirely different approach of a simplified contemporary background that also respected the "spiritual dedication."[25] The part of the pavilion where the *Pietà* would be exhibited was not a chapel, but Mielziner wanted to maintain an atmosphere of devotion for believers and nonbelievers alike.

Before the design, however, Mielziner needed to fully understand the safety requirements and determine, for the sake of efficiency, how to move an anticipated six thousand visitors per hour through the viewing area. He wanted the viewers to enter as though they were in a sanctuary, with a sense of peace and calm, "unruffled by the push and jam" that any fairground attraction produces. Since this was not realistic, he proposed a low-ceilinged anteroom of four rows or lanes through which single lines of visitors would pass. Here the atmosphere would be controlled—the lights would dim as the Gregorian chant was intoned. As viewers entered the main chamber, the aisle would gradually rise in height as they stepped onto a moving walk. Mielziner wanted to avoid the kind of rush to which museum visitors are so often subjected. He intended the speed of this walk to be as slow as he could get away with, which turned out to be twenty-five feet per minute on the slowest lane, and eighty on the fastest. The four lanes would be tiered, so that each would be slightly higher than the one in front, as rows of seats are in a theater, with the first lane only twenty feet from the sculptural group. The first three walks would be automated, but the last lane in the very back would not, so that those who wanted to view the sculpture in a more leisurely manner could do so.[26]

The problem that Mielziner and the architects struggled with was how to design the moving walk. Would it be a moving sidewalk with belts? The idea of a treadmill design powered not by electricity but by the walkers' tread was even floated, as impractical as it sounded. Another question was whether the walkway could be designed to curve. Airport baggage pickups had only recently introduced a radial treadmill that could accommodate a ninety-degree turn, but engineering consultants advised that it could not be adapted to public use. With safety as the primary concern, Frederick Voss, one of the architects, concluded that the straight-line conveyance with a continuous belt was the best approach to eliminate the possibility of accident.[27]

Mielziner would be able to control the height and angle of the installation to the viewer's best advantage—a consideration not possible at St. Peter's. In an interview, he recounted that one of the Vatican officials who had accompanied the statue on its journey to New York said, "We'll never be happy unless we do something like this for the *Pietà* when we get it back in Rome."[28] Since the only Vatican official on the ship was Vacchini, it was an acknowledgment that the Vatican knew that the placement of the *Pietà* on the chapel altar was not ideal. On October 21, 1962, the *Osservatore Romano* published an article by Dr. Redig De Campos, director of the restorational facilities of the Vatican Galleries and Museums, stating that the sculpture, removed from its plinth, had at some point been lowered, moved forward, and inclined almost six inches, or five degrees, toward the viewer to improve the perspective. The right side of the statue had been raised by three and a half inches, resulting in the figure of the Madonna being in an almost vertical position and the figure of Christ in a horizontal plane. "Thus the work assumed a false character—more balanced and perhaps better suited to its solemn surroundings—but far from Michelangelo's concept."[29]

Mielziner decided to make a full-size wooden mock-up of the oval base, on which he placed the Metropolitan Museum's plaster cast in a room large enough to accommodate three observation platforms. From here he could test the effectiveness of various tilts, both horizontal and vertical.[30] Mielziner chose blue for the background, not only because it is the color traditionally associated with Mary, but because the color's depth would complement the striking beauty of the white Carrara marble that Michelangelo favored. As a stage designer, Mielziner was an expert on color theory and avoided warm tones that did not recede as well. Another challenge for Mielziner was that the viewing chamber was rather wide. To avoid dwarfing the six-foot sculpture, he needed to frame it. For this, he designed a series of thin metal strips, arranged vertically in sets of twenty-four on each side of the pedestal, which would support a screen of four hundred flickering lights resembling votive candles, muted by a pane of blue glass. Something was still missing. Mielziner wrote that "since this *Pietà* is in essence part of the drama of the Descent of the Cross, I felt the need for a shadowy Cross, including draped Shrouds." This was very much in contrast to the brightly lit cross against the richly colored marble wall of the chapel setting at St. Peter's. Mielziner used a half-size scale model of the *Pietà* to set the scene.[31]

To prevent any damage to the *Pietà*, Cardinal Spellman insisted on a bulletproof glass encasement on all sides of the display. Knowing that this was not good news for Mielziner, Spellman wished to tell him in person.[32] On receiving the news, Mielziner immediately contacted Corning Glass vice president John Gates. Gates visited Mielziner's studio on March 13, 1963, to look over the model and discuss the problems involved. One option they considered was installing sheets of glass in metal mullions, but both agreed it would be aesthetically wrong. Spellman wanted bulletproof glass, but Gates explained that that kind of glass caused a distortion and was not optically acceptable for observing sculpture at some distance. Mielziner also consulted James Rorimer, director of the Metropolitan Museum, who agreed. Both men recommended tempered plate glass, which would be strong enough to prevent damage from any thrown objects or liquids.

Mielziner was also concerned about any danger of serious injury to visitors, if not to the *Pietà* itself, should the plate glass break. Gates recommended a new material that Corning had just developed, called Chemcord, which had a protective quality that plate glass did not. Furthermore, it was optically acceptable, with no distortion of color and just one-third the weight of heavy plate glass. Unfortunately, it was still too early in the development cycle to be available. Mielziner then turned to Pittsburgh Plate Glass Company and Owens Illinois to see if their engineers had a comparable material, but to no avail.[33] Using glass had its advantages—it was clear, could be easily cleaned, and was manufactured in large sheets; but it was heavy, could shatter, and it was not bulletproof. Plexiglas, or plastic, could be scratched and had a slight color but had the advantage of being bulletproof. Plexiglas could stop a .45-caliber bullet shot at maximum velocity, and if the sheet did shatter, the pieces would have smooth, not sharp, edges. In the end, Mielziner selected ceiling-to-floor (twenty-eight by eight feet), one-half-inch-thick Plexiglas panels, which weighed approximately seven hundred pounds each.[34] The Plexiglas was not without risk, however; once it was installed, the half-inch gap between the sheets would remain a vulnerable area.

As soon as the moving walkways and lighting catwalks were installed, the team experimented with the placement of the Metropolitan Museum's plaster cast and wooden base, adjustable both in tilt and height. These tests were completed before the permanent marble base was constructed.[35] A timetable had been established based on the assumption that preparations for Room 104, the *Pietà* display area, would be completed on schedule: Starting January 1, Mielziner would begin work on the installation above the suspended ceiling, including the catwalks and light bars. Two weeks later, he would be ready to install the acoustical ceiling concurrently with the sound system. Next would come the installation of all lighting equipment except the footlights. From February 5 and for the following month he would install the vertical rods and light strings, props, and vinyl and velour facings. Mid-March, the stage would be ready for the installation of footlights and the marble base, followed by the laying of the carpets. Mielziner planned a test run with the plaster cast of the *Pietà* from Metropolitan Museum of Art in place, so that he could adjust the lighting. By the end of March, he would oversee the final cleaning once the plaster cast was removed. Mielziner was then hoping to make all final adjustments to the lighting on April 1, which turned out to be overly optimistic. But knowing the possible pitfalls of his business, he had anticipated the delay by building in a margin of three extra weeks of preparation time.

Since the pedestal on which the *Pietà* sat in St. Peter's Basilica was not to be shipped to New York, Mielziner also had to provide a proper base. For that, he had hired Giuseppe Tommasi Studios, an ecclesiastical art and marble specialist, for the construction of a pedestal expressly for the pavilion display. Mielziner selected porphyry, a fine-grained purple granite, for the pedestal, at an estimated cost of $5,670.[36] The shipment was to be transported by the SS *President Hayes*, departing the port of Livorno (Leghorn), Italy, on February 18, 1964, and due to arrive March 2. Mielziner also designed a rug pattern to surround the pedestal, an adaptation of designs from early sketches of Michelangelo.

Mielziner contracted out the construction of the setting to Feller Scenery Studios of the Bronx. Peter Feller was a theatrical set builder who no doubt had collaborated with Mielziner on many an occasion. His firm also surfaced the interior walls of the pavilion with his thermoforming-vacuum machines. Feller seemingly supplied everything in the room but the *Pietà* itself. Velour for the drapery, carpeting, panels for behind the sculpture, the cross, chandelier, and railings for the treadmills—all this, along with painting and installation, came to just under $100,000.[37]

As soon as details of Mielziner's design became known, the criticism began, and the harshest came from the Catholic press. A *Denver Catholic Register* editorial complained that "the *Pietà* is to undergo the indignity of being 'staged.' . . . Someone seems to think a Hollywood touch must be added, here in America." The writer added that "if the sound of crashing marble is heard in the Vatican Pavilion one dark night, it may well be that the shade of Michelangelo is smashing his *Pietà*."[38]

Others were more enthusiastic. Irving Stone, the author of a recent biographical novel on Michelangelo, *The Agony and the Ecstasy*, congratulated Mielziner on his "absolutely stunning" design. He wrote that "the Chapel of the Kings of France was not built for the *Pietà*, and your setting will be. I think that's one of the immortalities of Michelangelo's work, that it can be placed in a deeply beautiful and totally modern setting, and not only lose none of its significance, but only gain."[39] Mielziner had his champions and his critics, but he would have to wait for the most important critics to weigh in—the public.

8

Moses Delivers

The Fair Opens

From the start, the White House supported the fair. President Kennedy attended the United States pavilion groundbreaking ceremonies on December 14, 1962, and declared, "This is going to be a chance for us in 1964 to show 70 million visitors—not only our countrymen here in the United States, but people from all over the world—what kind of a people we are. What kind of a country we have. What our people are like, and what we have done with our people. And what has gone in the past and what is coming in the future."[1]

The need for peace was never more urgent. In October 1962, the Cuban Missile Crisis had threatened nuclear disaster, and the US-Soviet space race intensified Cold War tension that had begun with the Soviet's 1957 launch of *Sputnik*, the world's first satellite. As the war accelerated in Vietnam, the growing civil rights movement and cultural shifts brought tumultuous change and anxiety to American life. The fair was a welcome balm. When Moses and the fair directors activated the countdown clock in the lobby of the administration building in Flushing Meadows on April 22,

1963, exactly one year before the fair's opening day, Moses invited the president to return in 1964 to open the exposition, but seven months later Kennedy was assassinated in Dallas, Texas. It was left to President Lyndon B. Johnson to open the fair.

The official opening ceremony was held at the newly built Singer Bowl, located on the fairgrounds and sponsored by the Singer Sewing Machine Company. President Johnson spoke to ten thousand invitation-only guests, while congregated outside the stadium gates were civil rights demonstrators shouting into bullhorns loud enough to disrupt Johnson's speech. Pockets of protests continued inside the fairgrounds. These were organized by the Congress of Racial Equality (CORE) to bring attention to racial injustice with nonviolent action. Two breakaway chapters took it a step further by organizing a highway "stall-in" in which protesters would stop their cars on the highway leading to the fair entrance, while others blocked entry to the subways headed to Flushing Meadows. When CORE had announced its plans a month earlier, it generated concern on all sides. On the grounds that this tactic inconvenienced the whole community, not specific targets, civil rights leaders such as Roy Wilkins, the head of the NAACP, expressed their disapproval at a meeting with newspaper editors, adding that it would be neither orderly nor nonviolent. Urging restraint was Attorney General Robert Kennedy, who believed the racial issue would be "the major domestic problem for a number of years to come." He also asked that newspapers downplay any disruptions.[2] What was at stake was the passing of the Civil Rights Act, which among other things would end discriminatory "Jim Crow" laws.

For his part, Moses would not tolerate protests of any kind, wanting to keep politics, whatever the stripe, out of the fair, for the simple reason that protests hurt sales. Protesters, both Black and white, were arrested, most were charged with disorderly conduct and, in some cases, trespassing, but by the afternoon the protests had been quelled, and visitors could continue along their merry way. News media reported that the "stall-in" ultimately failed; however, photographs suggested that the demonstrators' "sit-in" on the Triborough Bridge met with some measure of success. Police commissioner Michael J. Murphy sized up the episode concisely by describing it as "a day in which the President came to the world of fantasy and encountered the world of fact."[3]

Despite a tense opening, the fair lived up to its promise as a wonderland. Moses had been criticized for recycling the 1939 fairground plan, but he took the original design and improved it. He may have been inspired by Chicago's Columbian World Exposition in 1893, where planners located the fairgrounds on filled and reclaimed waterfront at Jackson Park. While making substantial infrastructure improvements, the 1893 fair committee built the Field Museum as a permanent structure for future generations. Chicago city planners used the same site for the Century of Progress Exposition of 1933, while extending waterfront development.[4] In New York this approach provided cost savings, though not as much as Moses anticipated, since much of the existing infrastructure had not fared well over the past twenty-five years. The underground electrical conduits could not withstand the marshy environment and had to be replaced. Where Moses departed from previous fairs was the decision to make all exhibitors build their own pavilions. Just as he had shifted the costs of his Title 1 projects to the site buyers, Moses shifted the construction costs from the Fair Corporation to the exhibitors—one more lesson learned from his experience during New York's 1939 fair.

As if to spite the BIE for its refusal to officially designate the fair as "universal," Moses employed the theme of the "universe" throughout much of the park. At the center of the fairgrounds was the twelve-story steel armillary, the "Unisphere," sponsored by US Steel and designed to stand as one of the few permanent fixtures in the park. It was installed on the same spot as the Perisphere and Trylon, the towering centerpiece of the 1939 fair; it was rumored that with the outbreak of World War II, the metal from those installations was melted down for war munitions. The three rings around the metal globe of the Unisphere commemorated the three satellites that had been sent up to orbit the Earth: Yuri Gagarin's *Vostok* spacecraft, John Glenn's *Friendship 7*, and the *Telstar* satellite.[5] Though critics dismissed it—too realistic—it was popular with the public and became the fair's space-age logo. Located opposite the Unisphere at the other end of the site was the Court of the Universe, where the Fountain of the Planets was the site of a sound and light show when the sun went down. Halfway down this corridor, which ran the length of the park, was the Court of the Astronauts, and on one side, the Court of the Moon, with the Lunar Fountain, and the other, the Court of the Sun, graced by the Solar Fountain. The most stunning of all the fountains, the Astral Fountain, was situated in the Court of the Stars.

The fairground was divided into five areas: Industrial, International, Federal and State, Transportation, and the Lake Amusement Area. Moses renamed the avenues but maintained the symmetrical layout. Of the five areas, the Lake Amusement Area was the smallest and, as its name suggests, consisted of a lake surrounded by entertainment venues and amusement park rides. The major pavilions in the Transportation Area were the Big Three automotive manufacturers—General Motors, Ford, and Chrysler—but it also included auto-parts companies, car rental, and gasoline companies. This very popular area was also where visitors could visit the US Space Park and the Underground Home, an ingenious answer to any threat of a nuclear attack, which never caught on.

Twenty-four states participated in the Federal and State Area, which was also home to the official United States pavilion, a stunning 150,000-square-foot building designed by Charles Luckman Associates. The two-story building, called "Challenge to Greatness," was supported by four columns and seemed to float above the ground. Its focus was on President Johnson's domestic agenda, coined the "Great Society" program. The most popular feature was a fourteen-minute ride, "The American Journey," in which moving grandstands, each carrying fifty-five passengers, passed by movie screens in a "you are there" experience of America's history from the arrival of Columbus to the launch of the space age.[6] Another popular exhibit, conceived by President Kennedy, was the "Hall of Presidents," featuring presidential artifacts, posthumously expanded to include Kennedy's rocking chair.[7]

Of all the states exhibiting, one of the most innovative pavilions belonged to Illinois. The organizers commissioned Walt Disney to develop "Great Moments with Mr. Lincoln," featuring a life-size "audio-animatronic" figure of Lincoln reciting excerpts from his speeches. Animatronics, as they came to be known, were robotic animations so lifelike that, in Lincoln's case, some thought they were seeing an actor play the part. Disney and his "Imagineers" were also responsible for three other exhibits: the Carousel of Progress for General Electric; the Magic Skyway for Ford; and Pepsi's It's a Small World, to benefit UNICEF. It was an opportunity for Disney to develop new attractions, paid for by corporations, and a test-run for the amusement park Disney wanted to develop on the East Coast. Ultimately Disney hired Moses's supervisor of construction, General William "Joe" Potter, straight from the fair to direct the construction of Walt Disney World, which essentially was Disney's extension of the fair itself.[8]

The Industrial Area, the largest of the zones, included participating Fortune 500 companies such as General Electric, American Express, Coca-Cola, Eastman Kodak, and other well-known names, plus some now long gone, including TWA and Bell Telephone. The message promoted was one of progress and the promise of a better life. The area also housed the twenty-thousand-seat Singer Bowl, the multipurpose stadium and site of the opening-day ceremonies.

The International Area occupied prime real estate and hosted eighty nations. Some countries unable to support their own pavilion collaborated with others to present national displays in a shared space, such the Central American pavilion, which represented five neighboring countries, and the African pavilion, a collaborative effort by twenty-four new republics that had recently gained independence from colonial rule. It was not lost on Charles Poletti, head of the fair's international section, that welcoming their participation would help ward off competing Soviet and Chinese influence.[9] This was also the section of the fairgrounds in which most of the religious pavilions were located.

Robert Moses did not publicly embrace any religion—though born Jewish, he had converted to Christianity through the Episcopal Church at some point after his years at Yale and Oxford.[10] He personally supported a strong faith presence at the fair, following the lead of philanthropist John D. Rockefeller Jr. and also William Church Osborne, a prominent civic leader who had sponsored the Temple of Religion at the 1939 World's Fair.[11] Moses wanted as many religious pavilions as possible. He had hoped for Jewish and Islamic representation, and although neither of these faiths had their own pavilion, they were represented. The American-Israel pavilion was sponsored by the American Jewish community, and Jordan's pavilion highlighted its Islamic heritage. Since Jordan was the home of many Christian holy sites, the pavilion was also designed to promote tourism to the Holy Land in a bid to attract Christian pilgrimages (this would all change after the 1967 Six-Day War). The pavilion featured the stained-glass *Fourteen Stations of the Cross*, designed by the Spanish painter Antonio Saura, and an exhibition of the Dead Sea Scrolls.[12] If it had been left at that, there would have been no problem. The pavilion, however, included a provocative wall-size mural, with text from the perspective of a Palestinian refugee child taking straight aim at the Israelis. The American-Israel pavilion and members of the Jewish American community, as well as Robert Wagner Jr., the mayor of New York, called for the removal of the mural. With no satisfaction from Moses, who had earlier banned protests at the fair, members of the American Jewish Congress and others peacefully picketed outside the pavilion without permits and were immediately arrested. By July the pickets were acquitted, but not before the dispute escalated to the Supreme Court of New York. The mural remained.[13]

Moses was not invested in promoting one religious group over another; ever the pragmatist, he also wanted to fill the lots originally intended for countries in the Soviet bloc and European nations who decided to boycott the fair. Moses remarked that "into the gap left by the departing Communists, the saints have come marching in."[14] In her study of religious pavilions at the fair, Julie Nicoletta documents the push to include religion early in the planning process. In a prescient observation, J. Anthony Panuch, vice president of Industrial, Federal, State, and Special Exhibits, proposed that if this fair were to contrast a free society with a communist one, "religion must be regarded as one of the principal institutions, which differentiates our way of life from a totalitarian society. As such, its place as an exhibit in the Fair seems clear."[15] Despite the lack of participation by the Soviets and the Eastern bloc, the threat of communism was never far from mind in the 1960s.

The *New York Times* noted that "it is difficult to tell where the fair begins and religion leaves off. Indeed, two traditionally inimical concepts—the worldly and the spiritual—seem inseparably combined here."[16] The Protestant and Orthodox Center was sponsored by the Protestant Council of New York and represented twenty churches and groups. For its dedication, Archbishop Iakovos, primate of the Greek Orthodox Church in North and South America and a president of the World Council of Churches, declared the fair was "the battleground for a new and concerted effort to overcome bigotry and division and serve God's people as God's servants."[17] One of that pavilion's featured exhibits was the charred cross from the Cathedral Church of St. Michael in Coventry, England, which had survived German bombing in World War II.[18] The pavilion included an auditorium that showed a twenty-two-minute allegorical film titled *Parable.* Written and directed by Rolf Forsberg, the film, with no dialogue, presented Jesus Christ as a clown and the world as a circus. Though the film received rave reviews from critics and was honored at the Cannes, Venice, and Edinburgh Film Festivals in 1966, many considered it sacrilegious, and the lack of visitors proved that the country was not yet ready for Jesus depicted as a mime-clown. Two pavilion organizers resigned in protest, and one visitor threatened to shoot up the screen if the film continued to be shown. The problem, aside from Christ as a clown, was that the film ended with a death but no resurrection.[19] For the most part, Moses did not interfere in a pavilion's choice of programming, but he made a rare exception in this case, asking the Protestant and Orthodox Center to not show the film the second season. The organizers, however, did not comply. Instead, they substituted a free-will offering at the end of each showing for the fifty-cent admission charge, and attendance actually increased.[20] Though controversial at the time, with the *New York Times* running no fewer than four articles on the work, it stands as a powerful film and is well worth the watch.[21]

The Church of Jesus Christ of Latter-day Saints pulled out all the stops for its $3 million pavilion, rivaling the Vatican's in cost. Not to be outdone by the *Pietà*, the pavilion included a nine-ton reproduction of the Danish sculptor Bertel Thorvaldsen's beloved sculpture, the *Christus*. A Mormon volunteer, Richard Eyre, recalled his own experience:

> I remember thinking how beautiful and lifelike it was. And candidly, the *Christus* is great because it is the resurrected Christ, but artistically, it can't hold a candle to *Pieta*.
>
> But one night, I was closing up the pavilion, and I thought we had everyone out. I was about to lock the door, and I noticed a man standing at the other end of the pavilion in front of the *Christus* statue. I hated to disturb him, but I put my hand on his shoulder from the back. He turned around, and it was then I realized he was a Catholic priest. [I] told him it was time to close the pavilion, and the tearful priest said "That's all right. I was just worshipping Christ."[22]

The Mormon Church built its pavilion with a three-quarter-size replica façade of the temple in Salt Lake City and acquired an adjacent lot, which was landscaped into a much-welcomed garden.[23]

The Pavilion of 2000 Tribes showcased the Wycliffe Bible Translators and their missionary work on four continents, and the Sermons from the

Science pavilion, sponsored by the Christian Life Convention of New York City, offered films and live demonstrations on the compatibility of faith and science.[24] The Russian Orthodox Greek Catholic pavilion was a replica of the Holy Trinity Chapel, which was built for the Russian colony on the northern coast of California in the early 1820s. Eclipsed by the Vatican pavilion's *Pietà* display, this pavilion showed the *Holy Ikon of the Virgin of Kazan*, circa 1400, formerly enshrined at the Kazan Cathedral in Moscow until the Bolshevik Revolution in 1917.[25]

High-profile architect Edward Durrell Stone designed the octagonal pavilion named for the famous evangelist, the Billy Graham pavilion. The Graham pavilion showed the twenty-eight-minute film *Man in the Fifth Dimension*, a reference to the dimension of the spiritual. At the Christian Science pavilion, visitors were introduced to the teachings of Christian Science and could access the latest news reports from correspondents of the *Christian Science Monitor*.[26]

Though not religious per se, and somewhat overlooked, Sudan's pavilion sponsored a contemporary Islamic-designed pavilion that offered one of the fair's priceless treasures. In addition to relics of its four-thousand-year-old Nubian civilization, it gave pride of place to the twelve-hundred-year-old fresco of the Madonna and child, which had only recently been unearthed, from a Coptic church near the banks of the Nile in northern Sudan. It was significant because it indicated the existence of Christianity in the region far earlier than previously recorded. While on display, the six-by-six-foot square icon began to show signs of deterioration, and access was closed to viewers for much of the fair.[27]

Both art and religion served to soften the hard edge of technology and added, in the eyes of one writer, "a more human-centered and spiritual perspective—not a denial of technology, but a call for reflection and introspection."[28] In his study of world's fairs, Arthur Molella observed that "the lack of a Russian pavilion prevented a replay of the US-Soviet confrontation at Brussels 1958. In fact the absence of such a major player amplified the effect of the blockbuster Pietà display."[29] The Brussels Fair had become a Cold War propaganda battle in which the Soviet Union sought to dominate. That tension was absent from the New York fair.

While relations between Moses and the faith communities were always respectful, the same could not be said of the New York arts community. The loan of the *Pietà* was not the only issue they had with Moses and the fair. They had lobbied for the formation of an executive committee of artists, architects, and critics who would offer aesthetic direction. This, in fact, was not a new idea but had been implemented at the Chicago World's Columbian Exposition of 1893. For the Chicago fair, the architect and city planner Daniel Burnham had assembled such a committee of leaders in the fields of architecture, sculpture, horticulture, and landscaping to develop and maintain a comprehensive and integrated neoclassical theme. As an innovator in architectural design and known for his impatience with traditional design, Louis Sullivan criticized the classical bias, stating that, "the damage wrought by the World's Fair will last for half a century from its date, if not longer." The more recent New York World's Fair of 1939–1940 extended an almost decade-long run of government-sponsored art patronage in the form of the New Deal work projects. That fair's Board of Design not only incorporated a broad program of art in the form of murals and sculpture but allotted a separate pavilion for both contemporary and fine arts.

In anticipation of a similar consideration, the Committee of Art Societies had arranged for an exhibition of contemporary American art with museum backing and financing for the 1964–1965 fair. What they had not anticipated was Moses's conservative reaction to modern art. Aside from the fact that he was repelled by much of the contemporary art he saw, Moses had two additional objections. First, he believed that an exhibition of contemporary art would court controversy and reflect negatively on the fair. He only needed to reference the reaction to art displayed at the Brussels fair six years earlier. Moses also worried that such an exhibition would not be commercially successful and recommended to the artists' committee that they recruit a corporate backer, such as Chase Manhattan Bank or American Airlines, but neither firm was interested.

Moses was also not interested in sculptural decoration, which was another break with past fairs. In her study of public sculpture, *Sculpture in Gotham: Art and Urban Renewal in New York City*, Michele H. Bogart examined Moses's uncomplicated reaction to modern art—he did not like it. When word got out as early as 1960 that there was to be a fair, sculptors began soliciting Moses for work. His priority, however, was to save money for the post-fair transformation of Flushing Meadows into a park, and as far as he was concerned, architectural sculpture was unnecessary. But there would be some sculpture in the fair, so Moses established a

Committee on Sculpture in 1961 to select work "from contemporary conservative to the more conservative avant-garde." Despite Moses's insistence to the museum directors that he genuinely welcomed their advice, he was only going to accept designs from artists he trusted—he was not going to take any chances on up-and-coming artists.[30] Two of the four sculptors selected had already worked for the 1939 fair, illustrating Moses's predilection for more traditional art. Marshall Fredericks's *Freedom of the Human Spirit* and Donald De Lue's *Rocket Thrower* were figurative and therefore deemed more conservative—too conservative for art critic John Canaday, who considered De Lue's work already out of style by 1939. Canaday labeled the *Rocket Thrower* the "most lamentable monster, making Walt Disney look like Leonardo da Vinci."[31] Moses got the last laugh, however, since he commissioned this and the four other sculptures as permanent park fixtures, on view for visitors to this day.

Moses enlisted another alumnus, Marshall Fredericks, whose 1939 contribution was the eccentric *Baboon Fountain.* For this fair, Fredericks chose a more compatible subject, *Freedom of the Human Spirit*, commissioned for the United States pavilion. The twenty-eight-foot bronze statue depicts a man and a woman soaring toward the sky, flanked by wild swans; its upward, graceful lines reflect a sense of optimism. Fredericks said "the idea was that these human beings, these people—us, do not have to be limited to the earth, to the ground. We can free ourselves mentally and spiritually whenever we want to, if we just try to do so."[32]

Moses did allow for two abstract works, both reflecting the fair's theme of space exploration. José De Rivera's *Form* was one of them. *Form* is a bow-like tapered ribbon of stainless steel that points upward as it slowly revolves on a pin mounted on a black granite base. De Rivera's sculptures were featured at both the 1939 New York and 1958 Brussels fairs, and De Rivera was one of the fortunate contemporary sculptors whose work had been acquired by the Metropolitan Museum's Department of American Art.[33] Canaday had approvingly praised his work as being "[as] perfect a fusion of technique and expression as contemporary sculpture—for that matter, sculpture of any period—offers."[34]

The second "abstract" work was Theodore Roszak's *Forms in Transit*, a massive, forty-three-foot-long sculpture of aluminum and steel tubes and sheet metal, which was more realistic than abstract and could be, with a little imagination, identified as a type of jet. Roszak, born in Poland in

1907, immigrated with his family to Chicago two years later and went on to study art at the National Design Academy and the Art Institute of Chicago. During World War II he worked on an assembly line at the Brewster Aircraft Corporation, and the influence of this experience is unmistakable in his work. He began to view the airplane with horror, as an airborne deliverer of mass destruction, but motion and flight would become his signature motifs.[35] He received recognition for the controversial thirty-seven-foot aluminum eagle he created for the exterior of the United States Embassy in London's Grosvenor Square in 1960. What would seem a handsome work today, a complement to the sleek façade of the Eero Saarinen building, was clobbered by segments of the British press, a debate that even made it to Parliament—it just did not fit in with the quiet dignity of the square, simply being in "poor taste."[36] Roszak was, after all, one of the "Irascibles," who only a decade earlier rejected Roland Redmond's invitation to compete in the Metropolitan Museum's acquisition drive. Time (and success) seemed to have mellowed his opposition to institutional work.

Moses's approach to art at the fair reveals that he regarded De Lue's *Rocket Thrower* to be the fair's centerpiece, as it referred to modern space exploration. The forty-three-foot bronze was the largest and most expensive of the official fair commissions. De Lue received $105,000, compared to Fredericks's $90,000. When it came to modern art, Moses and De Lue were in total agreement. "So much of modern art is political protest opposition to the established order," wrote De Lue to Moses, "and to my mind the visual evidence of the moral decay. That part that is difficult to understand is how great wealth such as the Rockeffellers [*sic*] and Fords can so heavily finance an ideology so dedicated to their own destruction."[37]

It was a Rockefeller—Governor Nelson Rockefeller—who with architect Philip Johnson invited pop artists to exhibit their work on the exterior walls of the fair's New York State Building. Government officials quickly objected to Andy Warhol's *13 Most Wanted Men*—installed one week before the fair's opening—which was a collection of mug shots of New York's most wanted criminals of 1962. The issue, at least the one offered by the governor's office, was that most of the men depicted were of Italian descent, and Rockefeller did not want to risk insulting this important sector of his electorate.[38] Warhol was told to either remove or replace the work in twenty-four hours. He instead chose to paint over the work with aluminum household paint. Although the pavilion successfully introduced the art of other pop artists, such as Roy Lichtenstein, Robert Indiana, and James Rosenquist, the incident with Warhol illustrated exactly the problem Moses sought to avoid.

The fair became a proxy in the battle for the soul of art: traditional representational art, embraced by the public, Moses, and Redmond, versus the art elites of contemporary abstract art. This was Moses's fair, but he was nothing if not pragmatic. Except for the Warhol fiasco, Moses left it up to each pavilion to sponsor its own art. Moses found a most unlikely ally in John Canaday, who in a go-ahead-and-shoot-me salvo, recognized the fair for what it was—a commercial venture. And if it happened to promote the nation's cultural life, so much the better. He wrote that "the people who will file past Michelangelo's *Pietà* may be stirred by religious associations and impressed that this really is by the man who got such a favorable write-up in *The Agony and the Ecstasy*. But unless they already know, this exposure is not going to make them understand what is great about the Pieta [*sic*] as a work of art." Hardly a ringing endorsement. Canaday was not particularly worked up over Moses's entirely "prophesiable" veto of a separate pavilion for contemporary artists.[39] Moses eventually relented and offered rent-free land one month before the fair opened, but the artists complained that there was not enough time to start building.[40] Moses was accused of "arbitrarily" killing off plans for an exhibition, but, owing to eleventh-hour financial difficulties, Argentina was unable to open its pavilion, making the building available for an exhibition of contemporary art. Though the Pavilion of Fine Arts, as it was called, opened a couple of months after the fair had already begun, its display of contemporary artists such as Milton Avery, Jasper Johns, Willem de Kooning, Andrew Wyeth, and Ben Shahn did manage to appease the critics (but not Canaday).[41]

This dispute reflected not only an unrelenting criticism of Moses's design choices, from the fairground's layout to its signature emblem, the twelve-story, stainless-steel globe known as the Unisphere, but the uproar over art signaled the growing cultural dissonance between the pre– and post–World War II generations, which would erupt in a few short years. Moses was already seventy-five years old, and he had septuagenarian tastes not only in art but also in music. His musician of choice was his friend Guy Lombardo.[42] Ten years earlier, Moses had awarded Lombardo the contract for musical productions at the Jones Beach Musical Theater. Now Lombardo and His Royal Canadians played six nights a week at the fairgrounds' Tiparillo band shell. In the meantime, Beatlemania had exploded in the US with the hit "I Want to Hold Your Hand." The group made their first American tour with a performance on the *Ed Sullivan Show* in February 1964, and as much as Moses hated "rock 'n' roll," he did eventually warm to the idea of a Beatles performance at the Tiparillo or the Singer Bowl. But by this time, as Joseph Tirella points out in his superb overview of the fair, *Tomorrow-Land*, the fairground venues would not have been large enough to hold them. When the Beatles returned in a year, they did perform in Flushing Meadows, but not at the fairgrounds—they played at Shea Stadium, another one of Moses's construction projects.[43]

Moses worked tirelessly to bring in additional attractions to sell the second season as "new." He had to—in a rookie accounting error, revenue from advance ticket sales for 1965 admissions had been spent in 1964; so to make his dream park a reality, Moses had to pull the proverbial rabbit out of the hat to best 1964's attendance of 27,148,280, which was still under the expected 40 million. Hoping to close the second season with a target attendance of 37.5 million, Moses spent $5 million on new exhibits, including a show of sixty modern painters and sculptors.[44] Pavilions also added new attractions to draw back visitors, such as the US pavilion's Hall of the Presidents, with memorabilia of thirteen presidents, including an original copy of the Bill of Rights and original drafts of Lincoln's Gettysburg Address. New also was the Churchill Center, a tribute to Sir Winston Churchill, opened by his daughter.[45] There is a reason why Moses called the fair a "Summer University," because these features still had drawing power.

1965: The Second Season

The fair reopened on April 21, 1965, with three times as many visitors as opening day the year before. They were greeted with a parade of eighteen bands, thirty-one floats, and four thousand marchers. Speaking at the Singer Bowl opening ceremony was the guest of honor, the mayor of Berlin, Willy Brandt, here to promote his city's technological and industrial strength. Since West Germany did not sponsor a pavilion, the West Berlin pavilion was clearly a BIE dodge. The real star of the day, however, was Vice President Hubert Humphrey, who gave the principal address at the ceremony. He received a rock-star welcome as he toured the fairgrounds in an open white convertible with a featured stop at the pavilion of his home state of Minnesota.[46] Before he returned to Washington, he paid a visit to the Vatican pavilion, accompanied by Chief Justice Earl Warren, the US secretary of commerce John F. Connor, and Cardinal Spellman. Two pavilion priests were guiding the group through the exhibits when Humphrey requested a pause for prayer. The six men stopped and prayed in the Good Shepherd Chapel on opening day.[47] Humphrey had a lot to pray about, as the *Times* reported on the front page of the day's edition that the White House was to "step-up" US involvement in Southeast Asia.[48] In a short while, combat troops were to follow. The full Americanization of the Vietnam War was on.

9

The Vatican Pavilion

Heaven on This Side of the Earth

The 1960s was a period of rapid change in the visual arts, reflecting the tumultuous cultural and political zeitgeist. The church was not immune from these trends and addressed the role of art at the Vatican Council, convening at the time in Rome. Cardinal Spellman expressed his reservations—too many changes, too fast. The council was looking not simply to modernize the Mass but to embrace and support the growing church throughout the non-Western world. Church art and architecture had not changed much from that of the nineteenth century. Spellman's mentor, Pope Pius XII, had not welcomed modern art and rejected the radical new church design ushered in by the Swiss-French architect Charles-Édouard Jeanneret, known by the moniker Le Corbusier. Many in the church did not consider modern art fitting for religious purposes. Art historian Jitka Jonová's study of sacred art during this period stated that "freedom of art presented a problem for the church."[1] The church did not claim any visual art style as its own, in contrast to the field of music, where the church owned Gregorian chant. Now faced with an entirely new wave of contemporary art, the council reexamined sacred art and its role in liturgical reform. The church had always been a supporter of beauty, of art, and granted freedom to various kinds of art, but also had to "ensure that the work of art was appropriate for the use within the liturgy."[2] Pope Paul VI, on the other hand, was open to modern art and demonstrated an interest from the beginning of his pontificate, considering this period as a "new springtime for religious art."[3] The topic was debated at the first session of the Vatican Council, which expressed its openness to contemporary art and issued the conclusion that "the art of our own days, coming from every race and region, shall also be given free scope in the Church, provided that it adorns the sacred buildings and holy rites with due reverence and honor."[4]

For his part, Cardinal Spellman gave the pavilion directors free rein to use art to present the church in its essential meaning—as Christ in the contemporary world. They balanced their approach with a combination of the rich artistic legacy of the past with current trends in abstract and pop art. The designers wanted to use this opportunity to educate and overcome the misconceptions of a non-Catholic public. The exhibits focused on the arc of Christianity from its Jewish heritage, the prophecy of the coming of the Messiah, through and up to the present day of Vatican Council II and its *aggiornamento*, a new openness to dialogue with the modern world. On the cusp of a seismic shift in the liturgical arts, the pavilion would both introduce and negotiate the challenge to the centuries-old ecclesiastical heritage.

In keeping with the pavilion's theme, the building's design was sleek and modern. Two key factors underpinned the oval-spiral design of the pavilion—the elliptical-shaped site of fifty-five thousand square feet, and the need to facilitate smooth crowd flow for an anticipated six thousand persons per hour on peak attendance days.[5] The exterior walls were finished in white cement stucco, mixed with marble chips to add sparkle. The pavilion was topped by the chapel, covered by a gilded roof in the shape of the *conopoeum*, a symbolic umbrella denoting the papacy, and crowned with a forty-two-foot-high cross. The cross was made of gold-anodized aluminum with stainless steel needles, which gave it a three-dimensional appearance, with glittering rays bursting from the golden center. With the addition of the cross, the pavilion exceeded the eighty-foot height limit, which was in place owing to the proximity of La Guardia Airport.

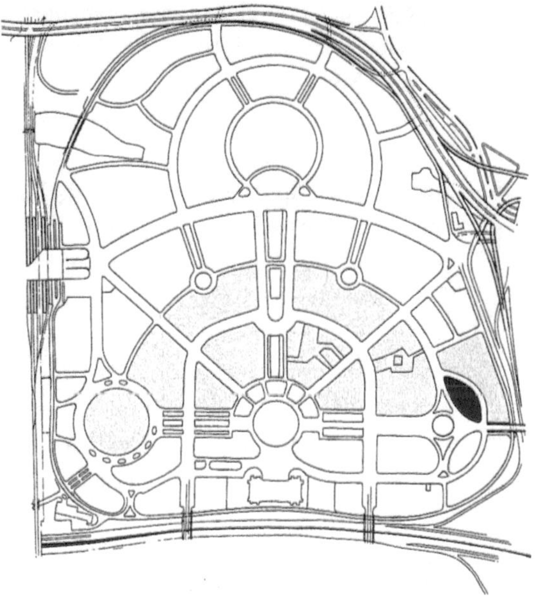

Figure 9.1. Map of fairgrounds, with the Vatican pavilion lot in black. (*Official Guide Book, Vatican Pavilion, New York World's Fair 1964–1965*)

However, with Moses's approval, the fair's Conformity Committee authorized this exception.[6]

A monumental court a hundred feet in length, featuring an impressive reproduction of the pontifical coat of arms set into the walkway, led to the pavilion's entrance. The adjoining sixty-foot-high wing wall included three built-in variable-height fountains to cool off visitors as they entered. Visitors also were greeted with specially selected music from two weatherproof Bozak sound columns mounted on either side of the balcony over the entrance, providing sound coverage for five thousand people.[7]

Visitors could also experience the fifty-bell Schulmerich "Americana" electronic carillon, loaned by the manufacturer, Schulmerich Carillons of Hatfield, Pennsylvania, for the duration of the fair. In addition to the loan, Schulmerich generously covered the cost of installation. The carillon was rung each day as the fair opened and was programmed to play automatically throughout the day. At noon and at 6:00 p.m. the bells marked time for the medieval prayer of the church, the Angelus, and played for Mass celebrations held inside the pavilion chapel. It was manually set for specially scheduled processions and religious celebrations.[8] Many of these selections, and more, were recorded by John Klein, the official carillonneur of the New York World's Fair, for the album *The Bells of the Vatican Pavilion*, produced by Americana Records.[9]

Although the Coca-Cola pavilion already had a Schulmerich 610-bell carillon, with an exclusivity clause, the addition of the Vatican pavilion's bells was cleared by Thomas Deegan, right-hand man to Robert Moses.[10] The Protestant and Orthodox pavilion, the Belgian Village, and the International Plaza also featured Schulmerich carillons, leading one to wonder how exclusive Coca-Cola's exclusivity clause really was—but the generous spirit of the company was not in doubt. Schulmerich also provided the Vatican pavilion with a full-scale replica of the four-foot-high, one-thousand-pound bell cast in commemoration of the historic Second Ecumenical Council that was concurrently taking place in Rome.

In addition to highlighting the contribution of the church throughout history, the pavilion was designed to provide a setting for the blockbuster of the fair, Michelangelo's *Pietà*. So that no one was denied this once-in-a lifetime opportunity to see this sculpture, the Rockefeller brothers, Nelson, Laurance, and David, contributed $100,000 to ensure that there would be no need to charge an admission fee.[11]

Pavilion Exterior—Area A

Like the decoration of Gothic cathedrals of the Middle Ages, sculpture on the exterior of the pavilion illustrated the teachings of the church, and for a contemporary interpretation the directors turned to Vytautas Kazimieras Jonynas, one of the best-known Lithuanian artists of the twentieth century. Jonynas came to the United States in 1950 and taught at the Catan-Rose Institute of Fine Arts in the city. With his former student Donald Shepherd, Jonynas founded the Jonynas and Shepherd Art Studio in New York specializing in church interiors, stained glass, sculpture, and murals. His works were prominent throughout the pavilion, including the cross atop the pavilion lantern dome.

Together Jonynas and Shepherd completed the central bas-relief and ten eleven-by-five-feet bas-relief sculptural panels decorating the pavilion exterior. In an interview, Donald Shepherd explained the selection of concrete for the panels, a not-uncommon technique at the time, in which concrete was poured into a mold. They chose this method "because it is contemporary by its very simplicity and represents the continuity of ancient Christian art."[12] The bas-reliefs were enriched with different colored quartz chips and mosaics. The panels served as a primer to the art of the church, beginning with *The Communion of the Saints* and moving counterclockwise from the pavilion entrance. Designed by Jonynas, the first panel was the largest of the bas-reliefs, twenty-seven by twenty feet, in the form of a dove representing the third person of the Trinity, the Holy Spirit. The subject was depicted with three elements: on the left side, the church militant, symbolized by the lamb, wheat, and grapes; on the right, the church suffering, with an angel freeing souls from the sufferings of purgatory; and in the center of the body of the dove, the eye of the Lord, the cross of the Savior, and the symbols of Saint Peter—the cross, the keys, and the rooster—representing the church triumphant.[13]

The next reliefs symbolized the *Creed of the Church*, beginning with *The Church—Promise and Fulfillment*, represented by the Ten Commandments and the star of David. The fulfillment is expressed in the symbols of the living church, a cross formed by a crosier (the bishop's staff) and a key, the symbol of Saint Peter and the office of the pope. In the center, the two Greek letters alpha and omega symbolize the beginning and the end, or Christ eternal. In the next relief, *God—the Holy Trinity*, the three persons of the Trinity are represented—the eye in the sunburst stands for God, the Father; the crucified Christ stands for God, the Son, and the dove, God, the Holy Spirit. Then came *The Prophecy of the Three Kings*, telling the story of Jesus's birth, depicting the Christ child in a manger, symbolically represented by the ancient Christogram, the Chi-Rho. The Chi-Rho are the first two letters of the Greek word for Christ, ΧΡΙΣΤΟΣ, chi (*X*) and rho (*R*) placed one over the other. The three kings are above, and surmounting them is a crown with its points pointing to heaven in fulfillment of the prophecy. Drawing on early church writings by Saint Ambrose, the fourth-century bishop of Milan, the *Blessed*

Mother is a lily, the roots are the Jewish nation, the stem is Mary, and the flower of Mary is Christ.

The next three panels represented the *Commandments of the Church*, beginning with *The Adoration of God. The Holiness of Marriage* is represented with a lily signifying purity and intertwined rings, and a man with a dove signifying that the human body is the temple of God, the dwelling of the Holy Spirit. The last three panels presented elements of the *Liturgy of the Church. The Holy Orders* are represented by the host, the chalice, and the stole as reminders of the priestly mission. In *The Sacrament of Baptism* the focus is on water, and just as Noah's ark was brought safely through the flood, now the water of baptism delivers Jesus's saving grace. *The Holy Mass* is signified by the sacrificial lamb of the Old Testament and the lamb of the Resurrection of the New Testament.

Together these panels represented the timeless sweep of Christian iconography, established in the earliest years of the church. The design team continued the rich artistic tradition throughout the interior, joining Renaissance genius with a contemporary vision.

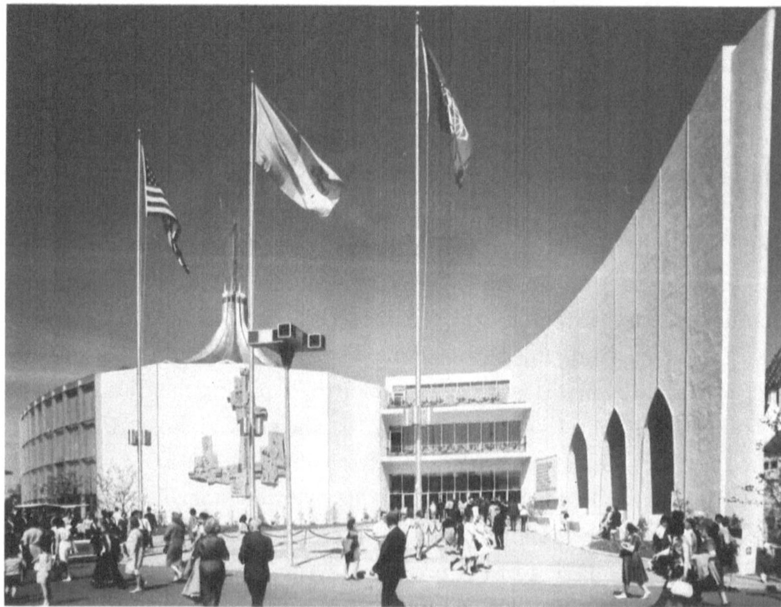

Figure 9.2. Pavilion exterior. (Photo courtesy of the Diocese of Brooklyn Archives)

Pavilion Interior

The pavilion was an elliptical, two-story building, with the second level extending only above the building's core. Like an elongated doughnut, at the core was the main Vatican exhibit, which featured a replica of the tomb of Saint Peter. The chapel was located directly above the tomb, echoing the placement of the altar at St. Peter's Basilica, built over the ancient burial site of the disciple of Christ and first leader of the church, Saint Peter. The pavilion was divided into several areas that wrapped around the core in a spiral. Each of the areas were themed, and these designations were helpful in locating, identifying, and understanding the individual exhibitions and the overall layout of the pavilion.

Once inside the pavilion, visitors found themselves enveloped by sound. A row of twelve hemispherical speakers were mounted every fourteen feet. The speakers were angled to project toward the ceiling, where the sound was reflected from a uniquely designed metal plate to provide an omnidirectional distribution pattern.[14] Early in the construction of the pavilion, R. T. "Rudy" Bozak, founder and president of the eponymous loudspeaker manufacturing firm, was the audio consultant. The success of the sound system was due to his careful planning and precision. A complex of seven distinct sound systems was installed; five provided taped

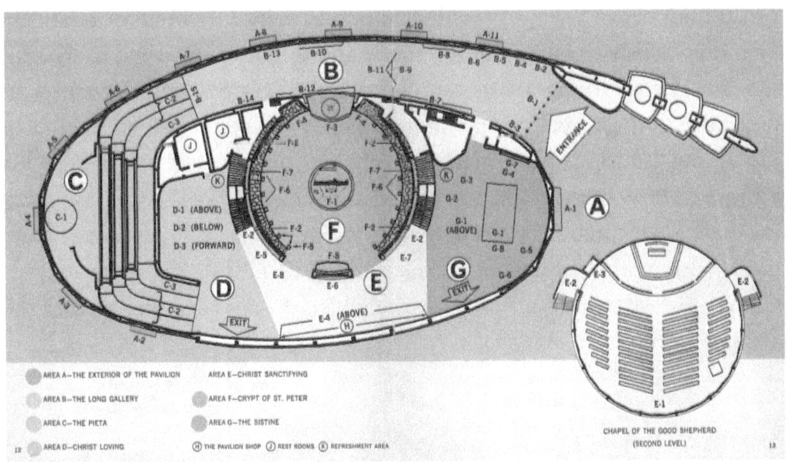

Figure 9.3. Pavilion floor plan divided in areas. (*Official Guide Book, Vatican Pavilion, New York World's Fair 1964–1965*)

material to various areas, one was for the chapel, and a seventh was an emergency public address system. Loudspeakers were selected according to the requirements of the various zones of the building.[15]

While the sound was referred to as "Gregorian Muzak" by detractors, others had only praise for the pavilion's music, which set the tone for the entire exhibit. Larry Zide, a columnist for the sound engineering trade magazine *Audio*, evaluated the sound systems of all the major pavilions, including Disney, IBM, Bell Telephone, and General Motors. After a half dozen visits to the fair, he returned for one final visit with his wife and children and concluded that the Vatican pavilion had "our vote for the best music sound at the Fair, and [we] joined the end of the line, long as always."[16] Visitors heard the sounds of Handel's *Messiah*, Bach's Cantatas nos. 4 and 10, "Jesu, Joy of Man's Desiring" (from no. 147), the duet from Cantata no. 78, and an aria from Bach's Christmas Oratorio. Also included was the finale of the Nativity Story by Heinrich Schütz. These were the sounds of the pavilion, and for the sights, visitors could follow the spiraling concourse, moving from one area to another, viewing the themed art and displays detailed below.

The Long Gallery—Area B

While the spotlight of the Vatican pavilion was on Renaissance art, world's fairs have always provided an opportunity for young or relatively unknown artists to launch their careers and showcase their work. The directors selected young artists representing the very best of midcentury liturgical paintings, sculpture, and graphic arts to display in an exhibit in the "Long Gallery." Narrowing slightly as it progressed, the gallery led visitors directly to the *Pietà* and served as an area of preparation for viewing the main attraction. Certain themes were emphasized and flowed from the Creation, the foretelling of the coming of Christ, the genealogy of Christ's ancestors, and his life and teachings. The narrative began with the ancient motif of the *manus dei*.

Mosaic of *Hand of God*

The "hand of God" is a motif in both Jewish and Christian art representing God or his intervention. Showing the full face or body of God was not yet permitted. Here the twelfth-century fresco of the triumphal arch of Saint Clement of Tahull at Barcelona is reproduced in a concave-shaped mosaic, measuring ten by ten feet and framed in brass. It was made by the Venetian Mosaic Works of Yonkers, New York. The surrounding walls were paneled in rosewood, with an acoustical ceiling, and the background music was from Handel's *Messiah*, selected to complement the theme of "prophecies," as were the following artworks.[17]

Stanley Bleifeld, *Prophecies of the Coming of the Messiah of the Old Testament*

Stanley Bleifeld's terra-cotta sculptures depicted prophets of the Old Testament, works that *Life* magazine described as "unpretentious drama and humanity." Well versed in the Old Testament, which he studied as a boy in Brooklyn, Bleifeld remarked that he was not religious but liked to "open the bible at random to get ideas."[18] Born in 1924, Bleifeld studied at the Tyler School of Fine Art at Temple University and earned his BFA and MFA. He lived and worked in Weston, Connecticut, until his death in 2011.

Bleifeld's commission developed from a chance meeting at a party attended by Ed Luders, the Stamford architect and design director for the Vatican pavilion. Luders was impressed by Bleifeld's work in Rome, *Il Credo*, which was exhibited at the Augustinian basilica of Santa Maria del Popolo, in the popular Piazza del Popolo, and ultimately earned Bleifeld the pavilion commission. Bleifeld, who was Jewish, was given "only the briefest guidelines to Roman Catholic ideas about the prophecies," but he studied them along with the commentaries, and finally chose five prophets to depict: Hosea, Daniel, Isaiah, Zacarias, and Jeremiah. Taken together, the five prophecies describe the Messiah as both a tragic and a heroic figure. Bleifeld sought to capture these ideas in his interpretations. In the bottom spokes of an eleven-by-seven-and-a-half-foot upside-down Y, he projects the prophecies of Hosea, with the Messiah as a leader appointed from the people; Daniel, the conquering warrior; Isaiah, the man of sorrows and the child with whom the lion and the lamb lie down; and Zachariah, the worthless shepherd, cast out by his people. Above them and separated is Bleifeld's interpretation of the prophecy of Jeremiah, one who "executed judgment and righteousness." Bleifeld saw the commission as a rare opportunity for an artist to compete with the engineering and architectural wonders of the world's fair.[19]

Emil Antonucci, *The Creation*

Emil Antonucci's contribution was an eye-catching multilingual depiction of the opening text of the book of Genesis. With pop-art-style graphics, the text was set in three-inch lettering in English, French, Italian, and Hebrew overlaying the background of multicolored Latin text: "In the beginning God created the heavens and the earth; the earth was waste and void. The darkness covered the abyss and the spirit of God was stirring above the water."

Today, the Brooklyn-born Antonucci is best remembered for his design of the logo for the New York hotel restaurant the Four Seasons, through his association with architect Philip Johnson. He was also the designer of the fair's New York State pavilion logo. But his real love as an illustrator and graphic arts designer was modern Catholic art. In a tribute to him, the American Institute of Graphic Arts wrote, "He sought to introduce Catholic art and graphics into the American mainstream."[20] He accomplished this as the illustrator for, among others, the Catholic magazine *Commonweal*.

Antonucci also contributed a three-dimensional presentation of the genealogy of Christ, from Adam to Jesse and David to Christ, *The Tree of Jesse*. The sixty twelve-by-nine-inch plaques were individually mounted on a swivel base for an interactive display.[21]

Doris Caesar, *The Annunciation*

Also born in Brooklyn, Doris Caesar began her sculptural work in the 1920s, turning away from classical forms and moving toward elongated forms for her secular and religious sculptures. The Whitney Museum had exhibited forty of her pieces in a four-person show titled *Four American Expressionists* in 1957. Her submission was a black bronze free-standing sculpture of an awestruck Mary, who had just received the Archangel Gabriel's message that she would conceive and become the mother of Christ, the event known as the Annunciation.

Fra Angelico, *The Annunciation*

The *Annunciation* was the most favored subject of Renaissance painting and the most beloved work of the Early Renaissance Italian painter and Dominican friar known as Fra Angelico, who painted this fresco for the convent of San Marco in Florence, where it remains to this day. Here a reproduction of the artwork was rendered in a full-color, back-lighted transparency. The use of transparencies or "illuminations" were quite popular at the fair, as photographic technology advances allowed for sharp reproductions of color images. Measuring six and a half feet square, it was positioned to lead the viewers to the Nativity. Its naïve and tender charm reflected Fra Angelico's simple and humble character.

Norman Laliberté, *The Incarnation*

Norman Laliberté's participation as a design consultant to the Vatican pavilion garnered him international recognition. One of his displays was a ten-by-sixty-foot photo-mural of a sixteenth-century German woodcut nativity scene, into which he inserted eleven light boxes. Through these light boxes visitors could view scenes of the nativity mysteries depicted in paintings selected and arranged by Laliberté.

Though born in Massachusetts in 1925, Laliberté grew up in Montreal. At the time he received the pavilion commission, he taught art at the Rhode Island School of Design. His use of bright colors and daring design and concept earned him the label of "radical." His art credo was "The abstract artist is not as meaningless as are the opinions of his critics. . . . If the artist is worthy and the effort of the viewer sincere he will be rewarded in having come one step closer to understanding God."[22] One either loved his new style of religious art or hated it. Judging by the reception of his work for the pavilion, visitors mostly loved it.

The Nativity

In a continuation of the narrative of the life of Christ, Laliberté created a display of crèches from various parts of the world and other Christmas trappings such as toys, candles, and angels. He presented what was traditional and familiar in an entirely unconventional exhibit.

Jean-Jacques Duval, *Stained-Glass Triptych*

The teachings of Christ were visually introduced through a stained-glass triptych designed by the French-born Jean-Jacques Duval. Duval made the piece in a style he developed known as *dalle de verre*, a thick, faceted glass created with a hammer and anvil rather than cut. Used as a space divider, the panels resembled the entrance to a medieval church.

Duval was born in 1930 in Strasbourg, France, to a Catholic mother and a Jewish father who died when Duval was just six. His mother never married his father, which as it turned out was fortunate, for during the German occupation the Gestapo tore their home apart looking for documents that would establish his Jewish paternity. After an uncle introduced him, when he was fourteen, to an artist who painted glass, glasswork became his lifelong passion. Duval studied art in Strasbourg, and it was here that a professor introduced him to liturgical stained-glass design. After completing his obligatory service in the French army, Duval came to the United States in 1952 and found work at the Daprato Glass Studio in New York. Drafted once again, he returned to New York after two years of service in the US Army.[23]

Ensconced in the Greenwich Village scene, Duval became friends with some of the leading artists there, such as Adolph Gottlieb and Willem de Kooning, pioneers in abstract art. By 1957 Duval had struck out on his own and established Duval Design Architectural Art Glass Studio. It is possible that Duval was first in the US to incorporate abstraction into ecclesiastical stained-glass designs, with a commission for a small convent on Long Island. As churches embraced modern design, there was a ready market for Duval's contemporary glass works.[24] Duval also created the award-winning stained-glass windows in the pavilion's Good Shepherd Chapel.

Norman Laliberté, *The Parables of Our Lord*

Retelling familiar biblical stories through the visual arts as a way of passing on the faith has a long history. Central to the teachings of Christ were the parables, seemingly simple stories that convey deep messages. Laliberté brought the fourteen parables of Christ to life by interpreting them in colorful, contemporary back-lighted montages and brought a more modern interpretation and sensibility to religious art.[25]

Donald Bolognese, *Miracles Performed by Christ*

In a series of eighteen 24-by-40-inch frames, Donald Bolognese displayed woodcuts in black and white and calligraphy titled *Miracles Performed by Christ*. Depictions of the images were accompanied by corresponding passages in calligraphy. The type of lettering Bolognese used was known as "chancery," the same style used for papal briefs and by members of Renaissance Italian society. Bolognese reproduced these images at one-third the size of those for a book he compiled by the same name, which was made available at the gift shop. Bolognese, who specialized in woodcuts, was a frequent contributor to *Jubilee Magazine* and an illustrator for numerous publishing houses. He taught calligraphy, book design, and illustration at the Pratt Institute in Brooklyn.[26]

Sister Mary Corita Kent, *The Beatitudes*

In a burst of vibrant rainbow colors, Sister Corita produced a forty-foot-long by five-foot-high serigraph (silkscreen) of the eight beatitudes, or blessings, often considered the New Testament parallel of the Old Testament Ten Commandments. This text formed the background for an interplay of quotations from Pope John XXIII and John F. Kennedy that reflected the message of peace and justice delivered by Jesus in the Sermon on the Mount.[27]

Kent was one of the more captivating artists contributing to the pavilion. A pop artist, she was passed over by the art history world—as art historian Susan Dackerman points out, she was not from New York, she was not a man, and she wore a habit. What distinguished her work from that of other pop artists was the social and spiritual lens through which she worked.[28] For this reason, it was her friend Norman Laliberté who commissioned her to create this banner for the Vatican pavilion.[29]

Kent was born in Fort Dodge, Iowa, and moved with her family to Los Angeles when she was five years old. She entered the Sisters of the Immaculate Heart in Los Angeles in 1936 and received her BA at Immaculate Heart College in 1941. There she began teaching art in 1947, and in 1951 she completed her master's degree in art history from the University of Southern California, where she specialized in serigraphy. Before she had her degree in hand, she won first prize in both the Los Angeles County

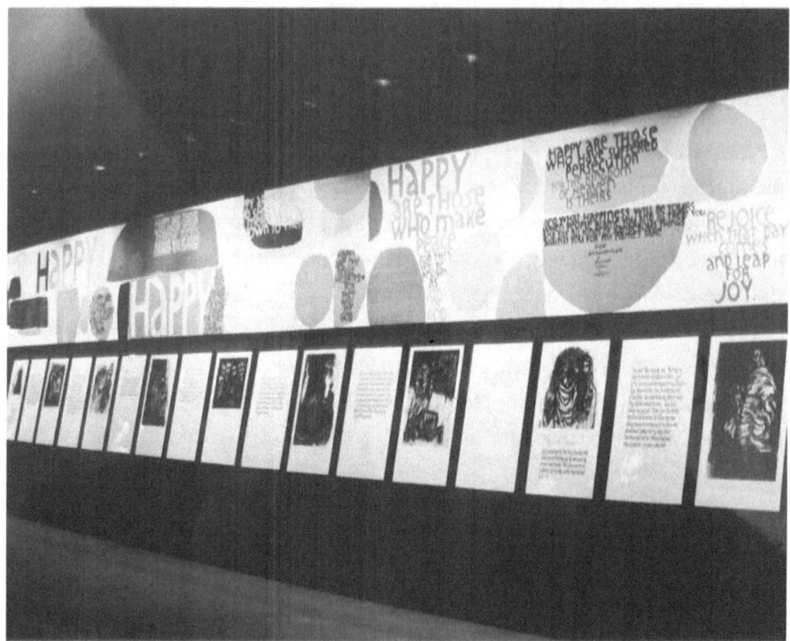

Figure 9.4. *The Beatitudes* by Sister Corita Kent (*upper band*) and *Miracles Performed by Christ* by Donald Bolognese. (Photo courtesy of the Diocese of Brooklyn Archives)

print competition and the California State Fair for *The Lord Is with Thee*. She became known as an iconoclastic artist, teacher, and activist.[30]

Kent initially leaned on religious images, with a medieval or Renaissance influence, but once she began to incorporate mundane objects with text from slogans, popular song lyrics, biblical verses, and literature, she saw that any art that was good had a religious quality.[31] Like other pop artists, she used consumer goods and imagery as inspiration. Her integration of these items into her pieces predated the practice by Andy Warhol, with works such as her early print *Wedding Banquet*, in which she depicts guests sitting on wire chairs designed by her mentor and friend Charles Eames. In another example, after a viewing of Warhol's *Campbell's Soup Cans* at his first solo pop art show at the Los Angeles Ferus Gallery in 1962, she used Wonder Bread wrapping to symbolize the Eucharist, while the polka dots resembled the Host.[32] Warhol was the biggest name in pop art at the time, but there were others who had a greater influence on her:

German art historian Alois Schardt, a professor and former director of the National Gallery in Berlin who was forced to flee Nazi Germany, and Charles Eames. Eames taught her that change was the "constant in art, and that each period really came out of the blood and bones and life of that time and couldn't be any other."[33]

Kent was not part of the mainstream art community—she was, at that time, too radical for the church and too Catholic for the art world. In an interview, she said that her biggest rejection came from Maurice Lavanoux, publisher of the *Liturgical Arts Quarterly*. As she recalled, "I had agreed to make a serigraph that fit in the magazine as a double-page spread so that each person who subscribed to the magazine would get an original serigraph. And I did 2,100 prints. . . . We sent them off to [Lavanoux] and didn't hear from him. So after a long time, I wrote and said, 'Did they get there?' And he wrote back, and he said he thought that it was a little too much ahead for his readers, so I said, 'Send them back.' "[34]

She found her own way through the social and political conflicts of the 1960s.[35] In 1968, she left the order, but not the church. She moved to Boston and continued her design work, receiving prestigious commissions from *Fortune 500* companies and the US Postal Service, for whom she designed the "Love" stamp in 1985. She was none too happy when the twenty-two-cent special was issued April 17, 1985, in Hollywood, California, with a dedication on the set for the TV series *The Love Boat*. Her influence was so far-reaching that a type font was named after her own calligraphic style, the "corita." In 2020, the French fashion house Chloé honored her in their spring collection titled "A Season in Hope," with five of her silkscreen artworks reproduced in their entirety on the garment fabrics. Licensed through the Kent estate, her work, a Chloé spokesperson said, complemented its "feminine spirit of optimistic rebellion."[36] Of all the contemporary artists who displayed work at the fair, it has been Kent's work that has been championed by a younger generation.

Perugino, *The Crucifixion*

Pietro Perugino was a leading painter in Renaissance Florence in the 1480s and was labeled the "best master in Italy" in 1500, whose most famous student was Raphael. Perugino's *Crucifixion* triptych, painted at the end of the fifteenth century for a chapel in the Dominican church in San

Gimignano near Siena, shows Christ hanging on the cross with Mary and Saint John the Evangelist at his feet. In the two side panels, Saint Jerome with his lion, and Mary Magdalene gaze up at the figure of Christ. The work does not attempt to depict the actual event or place but is a visual meditation on the theme of the Crucifixion. Here the panels were reproduced and displayed in transparency form, backlit for optimal viewing. The original is held at the National Gallery of Art in Washington, DC.

St. John the Evangelist, on the Deposition

Approaching the presentation of the *Pietà*, visitors could contemplate the text of Saint John the Evangelist's account of Christ's descent from the Cross (John 19:38–40), presented in a striking ten-by-twenty-four-foot panel with a woodcut roundel of Saint John. Here Joseph of Arimathea asks Pontius Pilate to let him take away the body of Jesus:

> After this, Joseph of Arimathea, secretly a disciple of Jesus for fear of the Jews, asked for his body. Nicodemus, the one who had first come to him at night, also came bringing a mixture of myrrh and aloes weighing about one hundred pounds. They took the body of Jesus and bound it with burial cloths along with the spices, according to the Jewish burial custom.

This verse was given special significance because it is believed that this is the passage from which Michelangelo received his inspiration for the *Pietà*.[37]

Above the entrance to the *Pietà* were three golden crosses, a reminder of the Crucifixion and the final preparation for the viewing of Michelangelo's sculpture of Jesus in the arms of his sorrowful mother. At the end of the gallery, visitors entered a softly lit passage, and the beautiful simplicity of the Gregorian chant helped build a sense of awe as they approached the *Pietà* (see chapter 10). From this area, visitors flowed into the next exhibit, on the contemporary church.

Christ Loving—Area D

After viewing the *Pietà*, visitors entered the gallery dedicated to Christ Loving, through representations of the church's love for humankind, as evidenced through the religious, educational, and charitable work at home and abroad. Images were projected simultaneously on ten screens, depicting scenes of charities feeding the hungry, clothing the naked, and caring for the sick, the aged, and the handicapped. The work of missionaries who spread the Gospel throughout the world was included, as well as images illustrating the role the church plays in education. Another wall displayed statistics relating to the worldwide activities of the church, presented in a scroll-like manner, highlighting its universal scope.

The Children's Corner

A special children's corner was outfitted with educational and multimedia amusements, such as View-Master-like light boxes, three-dimensional devotional scenes, and two screens that showed animated films and cartoons of Bible stories. Another wall depicted a medieval village with a collection of crosses from around the world, designed by Norman Laliberté. Attractions to teach the faith included a Calendar of the Saints, a Litany of Mary wheel/carousel, a Jesse Tree umbrella, and a reliquary doll. Sure to draw attention were a machine that made plastic models of the pavilion itself and a castle-like vending machine that would dispense a holy card for a penny.

The Wall of Information

A curved wall contained a compendium of worldwide church news reports presented in checkerboard squares of people, places, and quotes from those impacting the global community. From here visitors moved into the last of the three galleries.

Christ Sanctifying—Area E

The core of the Christ Sanctifying Gallery was an emphasis on the church as primarily and ultimately an instrument of sanctification—that the church's regard is for the whole human, "here and hereafter," and that it is from Christ through Saint Peter and his successors that the church receives its sanctifying power. This was the only two-level area, with the crypt of Saint Peter below and the Chapel of the Good Shepherd above. A wide, curved staircase led to the mezzanine level's circular chapel, designed as if to crown the pavilion itself. As the devotional center of the pavilion, the chapel was its spiritual heart, with seating for 350 and open all hours the pavilion was open.

The chapel was named after the ancient statue the *Good Shepherd*, on loan from the Lateran Museum of the Vatican. Eclipsed by the *Pietà*, the *Good Shepherd* could well be the more historically important. No more than three feet tall, it has been dated to the third century and is one of the first sculptural depictions of Christ, made during the persecution of Christians under the Roman Empire. The theme of a young shepherd boy was already a common motif in Roman-Greco art, such as the *Kriophoros*, a depiction of a young god carrying a ram on his shoulders. This image was easily appropriated by Christians, interpreted as Christ in search of the lost soul, a reference made several times in the Bible.[38] Until Christianity

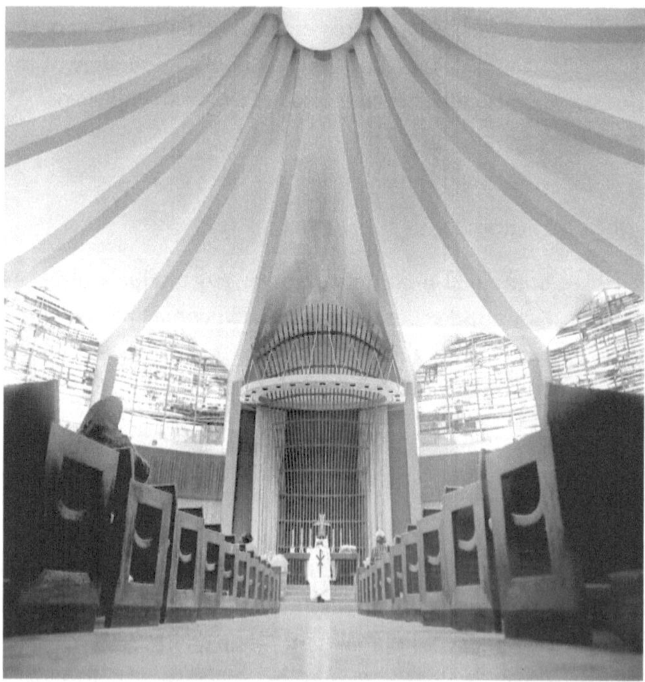

Figure 9.5. Good Shepherd Chapel interior. (Photo courtesy of Frank H. Methe III)

was legalized in the fourth century, such depictions were found only in the burial crypts, the catacombs. The statue on display in the chapel was the finest of them all. It was discovered in the early nineteenth century, broken into numerous pieces and missing the legs below the knees. Restoration work replaced the legs, the fingers of the right hand, and the head of the lamb. The sculpture was acquired by the Lateran Museum of Christian Antiquities, now part of the Vatican Museums. Originally the statue was positioned in a recess close to the pavilion entry, and visitors missed it as they entered. It was repositioned for the fair's second season, on an elevated pedestal behind the altar, to be seen by all.

Visitors to the chapel could take in the latest liturgical trends, reflecting the modern aesthetic of the day. Stained-glass windows designed and manufactured by Jean-Jacques Duval encircled the chapel. The thirteen windows at ceiling level covered twenty-four hundred square feet and were made of handblown, colored antique glass. They reflected Duval's abstract expression of Christ, King of the Universe, and the mysteries of eternal life. Nine windows at a lower level were Duval's interpretation of the Stations of the Cross. These were not typical stations that realistically depicted Christ's ascent to Calvary, his Crucifixion, and his entombment, but rather an interpretation in color and glass. Intense blues symbolized death, and small yellow areas represented the new hope that Christ brought. Above all, Duval wanted to create "an atmosphere of meditation, prayer and spiritual beauty for all who come to worship in the chapel."[39] Some people loved it, others did not.

The chapel table altar, designed by V. K. Jonynas, was made of wood and cast aluminum to resemble a dove in flight. Centered on the face of the altar was the ancient symbol for Christ, the Chi-Rho, and superimposed on the Christogram was a fish, which when written in ancient Greek as ΙΧΘΥΣ forms an acrostic for Ιησοθς Χπριστος Θεου Υιος Σωτηρ, which translates as, "Jesus Christ, Son of God [Our Savior]." The symbol was adopted by early Christians as code to identify themselves one to the other. At the tip of the fish's mouth was a small round wafer, the Host, symbolic of the Holy Eucharist.

The walnut reredos, or altar backdrop, swept up from behind the altar to form a type of baldachino, or canopy placed over the altar. Long drapes could be drawn to conceal the altar for events included in the pavilion's daily program, such as concerts or lectures. Missing was a confessional, allowing room for the organ.[40]

From the close of the first season to the opening of the second, the Vatican Council's influence was felt in various ways. The pavilion directors were receptive to new developments emerging from the council, which included how Mass was to be celebrated in the Good Shepherd Chapel. The question arose as to whether, in line with the "new liturgy," the altar table could be moved forward to allow the priest to offer Mass facing the people. The directors agreed that it could, since visitors would be expecting the new liturgy at this pavilion. Changes such as these could only be made with the bishop's sanction. Cardinal Spellman and Bishop McEntegart approved.[41] Also, the new liturgy was to be used at all Low Masses, and the congregation present was to participate. This meant the use of English instead of Latin, unless otherwise requested by the priest celebrant.[42] Mass was offered at various times every day, often celebrated by visiting priests from around the world, including those of the Eastern Rite. The chapel also hosted concerts and organ recitals for the weary faithful.

The area directly below the chapel, Christ Teaching (Area E), on the ground level, featured an unexpected star of the show: Norman Laliberté's liturgical banners. Prior to World War II, church art relied on traditional design, particularly fine needlework completed by artisan guilds. After the war, contemporary styles began to emerge from new directions in theology, with themes of celebration, joy, and thanksgiving. The Second Vatican Council "formed a watershed in liturgical style in North America."[43] Suspended from the pavilion ceiling were forty-four double-sided banners, each fifteen by five feet, visible from both the chapel and the main concourse. Laliberté blended ancient tradition with contemporary design to depict milestones in the church's historic calendar. On one side was the temporal cycle, the great feast days of the liturgical year, and on the other, the dominical cycle, a complex system of determining in advance the day of the week for any date, based on the date of the first Sunday of the year.[44] The banners were either silkscreen on linen canvas or specially woven multicolored fabric with appliqués of velvet, felt, and satin. They were a bold yet attractive way to decorate the glass part of the building with virtually no solid wall surfaces. Each banner hung from a horizontal pole and could be removed for special feast day processions. The display's

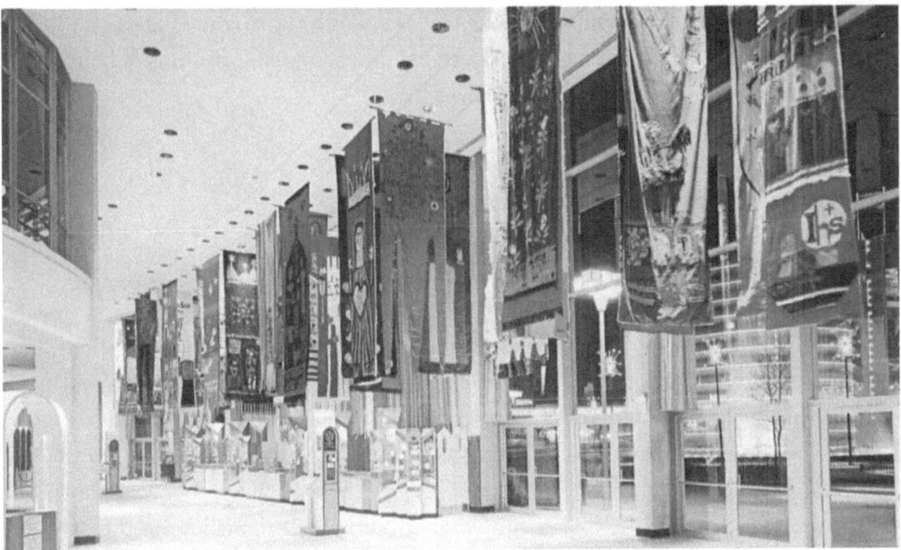

Figure 9.6. View of the pavilion's main concourse and the gift shop, with Norman Laliberté's banners. (*Official Guide Book, Vatican Pavilion, New York World's Fair 1964–1965*)

success may have contributed to the popularity of liturgical banners, which only now has begun to recede.[45] Laliberté brought a level of sophistication to this art form that has not been replicated since.

As visitors continued along the concourse, special honor was given to the three churches most significant to the fair and to Catholics in America: St. Patrick's Cathedral in New York City; St. Peter's Basilica in Rome, the heart of Christendom; and the National Shrine of the Immaculate Conception in Washington, DC. The churches were depicted in lead bas-relief, and beside the reliefs were passages specially selected to highlight their significance. Next to this display was the Catholic information booth for those with questions about the Catholic faith.

Seventy priests volunteered once a month to staff the information booth. The priest-volunteers came from the Archdiocese of New York, the Diocese of Brooklyn, and from twenty-one religious orders, and were there to answer questions seven days a week, twelve hours a day, every day of the fair. Their no-pressure approach found favor with visitors—one non-Catholic commented that "since I entered this Vatican Pavilion

I have been on my own. No one has intruded into the privacy I desire and like. You are here. The sign says I can ask questions. But nothing is being rammed into my heretical hide. This I like."[46]

Crypt of Saint Peter—Area F

Replica of the Shrine-Tomb of Saint Peter

The corridor of exhibits wound around the central point of the pavilion: the replica of the shrine-tomb of Saint Peter. The 1940–1950 Vatican excavations under the main altar of the basilica uncovered what is believed to be the burial site of Saint Peter. A faithful reconstruction of the tomb was built in Rome by the studio of Giorgio Leonori and shipped to New York. This replica was based on what could be determined from the excavated tomb, dated to shortly after the middle of the second century, around 160 AD, and referenced by the Roman Gaius, who wrote, "And I can show the trophies [tombs] of the apostles. For if you choose to go to the Vatican or to the Ostian Way you will find the trophies of those who founded this Church."[47]

Peter was the apostle Christ selected to serve as the first leader of the early church. Persecuted by the Romans, he was martyred by inverted crucifixion (nailed to a cross upside down) in Nero's Circus and buried not far away. It was on this spot, recognized as holy ground, that early Christians venerated and built his shrine. Over this shrine-tomb, Emperor Constantine built St. Peter's in the fourth century. The original structure gave way to the "new" St. Peter's in the late sixteenth to early seventeenth centuries. Modern excavations from 1939 to 1949 revealed a small monument built into a niche in the wall—this is the tomb that was reconstructed for visitors to the fair.

Doctors of the Church

Behind a circular colonnade and framing the crypt area were panels with representations of the "Doctors of the Church," along with a quotation of their teachings. A doctor of the church is the title given by the church to a man or a woman in recognition of their significant contribution to theology or doctrine through their research or writing. The panels, beautifully produced by artisans in Rome, leaned on the Byzantine style, beginning from the left of the crypt entrance and continuing clockwise. The idea was to illustrate the continuity from the early church to the present, leading to the worldwide meeting of bishops in Rome.

Second Vatican Council

A huge photo-mural of the Second Vatican Council convened by Pope John XXIII illustrated the monumental event that was held concurrently with the fair. Many of the visitors to the pavilion followed the sessions and were able to track the almost three thousand bishops, cardinals, and heads of religious orders by a projection of the world, with lights highlighting their countries of origin. It was at these historic sessions that the church authorized significant changes in the celebration of the Mass, which allowed the service be said in the vernacular instead of Latin. The council promoted a more active participation of the laity and a more ecumenical approach to other religions. Many other changes followed, including the call of religious orders to adapt to the modern world and the lifting of centuries-old traditions, such as relaxing the abstention from meat on Friday.

Mosaics

To the left and right of the photo-mural were simulated mosaics of the two Vatican pavilion popes, with accompanying text from their teachings. On the left was Pope John XXIII, with an excerpt from his encyclical letter "Toward a New Pentecost," and on the right was the image of Pope Paul VI, with a quote from his reopening of the Vatican Council in September 1963.

Columns of the Centuries of Christianity

An innovative approach to tracing the arc of Christianity was through the architectural feature of the "column capitals." Twenty columns encircled the crypt area, and atop each column was a capital re-created in the architectural style of each respective century. It focused primarily on church buildings in Italy and other Mediterranean countries.

Biblical Writings and Sacred Scriptures

Priceless manuscripts and documents from the twelfth to the sixteenth centuries were on display in glass cases. In the spirit of ecumenism, the General Theological Seminary graciously loaned its priceless copy of the Gutenberg Bible, dated 1455. Insured for $500,000, it was one of only forty-six printed Gutenberg Bibles still in existence, one of the rarer copies printed on vellum, and one of only six complete Gutenbergs in the United States. The Pierpont Morgan Library, the Union Theological Seminary, the Jewish Theological Seminary, and the New York Public Library also lent rare manuscripts for exhibit.

Teachings of Pope John XXIII

Special attention was given to Pope John XXIII, without whom the pavilion would not have been built. His legacy was marked in part by his landmark teachings on peace and justice. Engraved on wall panels on either side of the crypt area were quotations from two of his encyclicals, including one on the dignity of man:

> **The Dignity and Worth of Every Human Being**
>
> A human being is a person endowed with intelligence and free will. He is redeemed by the blood of Jesus Christ. He is by grace a child and friend of God and heir of eternal glory. He has the right to life, to bodily integrity, to respect for his person, to honor God according to the dictates of his own conscience. He has the right to work, also to go about his work without coercion. He has the right to private property, even of protective goods.

Jesus Christ, the Teacher

Leaving the crypt area, visitors could view an eight-and-a-half-by-four-foot plaster bas-relief of *Christ amid the Apostles*, reproduced from a fourth-century sarcophagus in the Church of St. Ambrose in Milan. The sarcophagus is one of the most famous examples of early Christian art.

To keep the flow of visitors moving throughout the various sections, the Vatican pavilion, like the other seven religious pavilions, depended on an army of volunteers, who served as hostesses, guides, counselors, and sometimes cleanup crews. Speaking to the *New York Times*, Rev. Joseph T. Lahey, the associate director of the Vatican pavilion, explained that they relied on forty volunteers every day. "We didn't think we'd have a need for too many when the fair first opened, but now we see we couldn't get along without them. They're a Godsend, very faithful."[48]

The Sistine—Area G

On display in the final exhibit area were the transparencies of Michelangelo's *Last Judgment* and frescoes on the Vatican's Sistine Chapel vaults, a fitting complement to the *Pietà*. Produced by *Life* magazine (and no relation to the Hollywood movie set), the exhibit had its genesis in one of a series of photo essays to bring art, culture, and history to the magazine's readership. Only after World War II was *Life* able to bring color cameras into Europe, and in 1949 the editors planned a special issue dedicated to the Sistine Chapel. Its December 26, 1949, cover story on the Sistine ceiling turned out to be the most popular of all its art features throughout the magazine's thirty-six years of weekly publication, selling out immediately. It included an elaborate full-color foldout, allowing readers to see the ceiling's entire length. For exhibition purposes, the transparencies from these photos were enlarged section by section and applied on a single sheet of translucent plastic, then mounted in light boxes for an exhibit at the Metropolitan Museum of Art in November 1956. Their installation at the fair made the Vatican pavilion the closest experience visitors could have to being at St. Peter's without leaving the country. The reproductions were the same size as the original reproductions, except for the Sistine ceiling, which was reduced slightly to fit the museum's exhibition space.[49]

Numismatic and Philatelic Display

The pavilion contained something for everyone, including a stamp and coin collection, which highlighted popes and presidents, on loan from Cardinal Spellman. Spellman once wrote that "stamps are miniature documents of human history. They are the means by which a country gives sensible expression to its hopes and needs; its beliefs and ideas."[50] Critics of Spellman found it to be an easy target for ridicule, but for stamp lovers it was of interest. So vast was the collection that it became the core of what is today the Spellman Museum of Stamps and Postal History, located on the campus of Regis College in Weston, Massachusetts.

News of the Contemporary Church

News releases of current activities of the Catholic Church, updated frequently, were available and included the latest reports from the Second Vatican Council.

Views of the Contemporary Church

A montage of the contemporary church was presented by rotating backlit color slides projected from seven slide carousels, the latest film technology from Eastman Kodak. These included views of traditional churches, the Mass and the sacraments, Pope Paul VI and the Vatican Council, ecclesiastical manuscripts, the Eastern Church, the contemporary church, and attractions of the Vatican.

The Church in the World

"Church Is Christ Alive in the World" was the title of Thomas Merton's script for a film that was shown at frequent intervals, which commented on the theme of the pavilion. Merton was one of the most influential Catholic writers of the twentieth century, the author of the best-selling *The Seven Storey Mountain*, his search for peace and faith. He was also a theologian, mystic, poet, and Trappist monk.

The Church and the Saints in Contemporary Art

Part of the Sistine Area was reserved as an art gallery, with changing exhibits that presented the works of contemporary artists such as William G. Congden, André Girard, and Andrés Salgó. The pieces, which included paintings, sculpture, and other media, expressed the theme of the saints. Hardly recognized today, the artists were known and respected in the post–World War II art milieu.

The American-born artist William G. Congdon (1912–1998) returned from World War II to New York, where he befriended Betty Parsons, who gave him his first one-man show at her eponymously named gallery. His paintings were shown alongside the work of friends and fellow abstract expressionist artists Mark Rothko and Clyfford Still. While the art of Rothko has exploded in popularity, Congdon's has not. In a review of Congdon's life and art, Giuseppe Barbieri writes that the artist committed professional suicide when he went public with his 1959 conversion to Roman Catholicism, losing the following of critics and the public.[51] The tradeoff, however, was worth it, as Congdon himself wrote in his 1961 account, *In My Disc of Gold*: "Conversion reinstated me as an artist, not of any particular kind, religious or secular, but simply as the artist the Lord intended me to be."[52]

Another featured artist, André Girard, born in France, studied under Georges Roualt and Pierre Bonnard and designed posters until World War II. He established the underground resistance Carte network, eventually fled to England, and then moved to the United States, where he received the Legion of Merit for his wartime efforts. After the war, he resumed his artwork, focusing on sacred art in the 1950s.

Andrés Salgó, of Mexico, was an artist in his country's muralist tradition, who painted a series of ten large oil paintings titled *Message of Peace*. The vibrant works expressed the spiritual hopes of humanity and the universality of religion, one of the central themes of the Ecumenical Council. Salgó was also known for his mural *Alliance for Progress*, a tribute to President Kennedy, at the main library of the University of Puerto Rico.[53]

More contemporary art was displayed along the lower section of the niche-wall, with a permanent exhibition of art on loan from St. Joseph's Abbey of Collegeville, Minnesota. Contrary to criticism that the pavilion was ensconced in the past, these efforts showcased modern design as part of a continuum of religious art through the ages.

As was often the case, the sharpest critiques of the pavilion came from Catholics, as if the pavilion was a proxy for Cardinal Spellman himself. A Catholic priest of the Carmelite Order, Gregory Smith, wrote a critical article in the order's own publication, the *Scapular*, that was picked up by the *New York Times* and probably got more coverage than it deserved. He criticized the decision not to hire an architectural superstar like Eero Saarinen, who designed the IBM pavilion, or Edward Durell Stone, who designed the Billy Graham pavilion. The response would be, of course, the cost! He acknowledged that there was "a dignity in the composition of its structure, even a certain elegance. But architecturally," he continued without elaborating, "what the Vatican Pavilion says has all been said before." Smith took aim directly at Cardinal Spellman, questioning the rationale of including his Vatican stamp and coin collection. He panned the scale model of Vatican excavations of Saint Peter's shrine-tomb as "an archaeological hangover or an antiquarian oddity somehow misplaced in the space-age." Perhaps his most egregious, almost sacrilegious complaint was questioning the selection of the *Pietà*, which in his mind bore little "religious relevance in the twentieth century." In contrast, Smith praised the welded-aluminum sculpture of the *Risen Christ* that towered over the entrance to the Vatican pavilion at the 1958 Brussels World's Fair, as "it spoke the language of today's world." He acknowledged that that sculpture had probably "fallen into some post-fair limbo of oblivion that the *Pietà* will never know."[54]

10

The Agony and the Ecstasy

Michelangelo Meets Hollywood

The *Pietà* had huge drawing power, not only because of its spiritual beauty, but also because of the widespread success of Irving Stone's historical novel *The Agony and the Ecstasy*, published in early 1961, which was quickly featured in the Book of the Month Club. Stone already had tremendous success with his earlier novel on the life of Dutch painter Vincent Van Gogh, *Lust for Life*, which was made into a movie in 1956. Writing with an eye on Hollywood, and a license to invent, Stone filled historical gaps and imagined dialogue based on his four years of research in Florence and Rome. His preparation included apprenticing himself to a sculptor to immerse himself in what the artist experienced. He also hired Italian students to translate all 495 of Michelangelo's letters from Italian into English. Stone compiled these letters and spun off a second book, *I, Michelangelo, Sculptor: An Autobiography through Letters*, the following year.

Hollywood immediately took notice, and by 1965 the movie adaptation was released, focusing on Michelangelo's explosive relationship with Pope

Julius II during the painting of the Sistine Chapel ceiling. In the novel-based movie, Charlton Heston starred as Michelangelo, and Rex Harrison played Julius II. Doubleday Books suggested that Stone prepare yet a third book, a fifteen-thousand-word edition just on the *Pietà*, priced at an affordable dollar to coincide with the run of the fair—publicity an author could only dream of.

Twentieth Century Fox offered the Sistine Chapel set used in the film production of *The Agony and the Ecstasy* to supplement the Vatican pavilion. Moses suggested that it could be installed at the Garden of Meditation, about two hundred feet away from the pavilion. The studio offered to underwrite the cost of shipping, but Moses advised that the Vatican pavilion would have to charge an entrance fee to cover any studio out-of-pocket expenses. All profits would accrue to the pavilion, but Spellman passed on the offer. He had no objection, however, to it being displayed at the Hollywood pavilion, as long as there was no mention of the Vatican pavilion.[1] It was too much commercial razzmatazz for such a holy exhibit.

The *Pietà* Area

Simplicity was the order of the day in the *Pietà* area of the pavilion, dedicated completely and solely to spotlight the sculpture. Jo Mielziner's intention was to offer visitors a completely new way to experience the sculpture in a contemporary yet solemn context. All the carefully placed art the visitor viewed, and the music heard in the approach, was to build the spiritual drama. After making their way through the Long Gallery, visitors entered the anteroom, with dimmed lighting and softer music. The atmosphere of this low-ceilinged, low-lighted interior chamber prepared the visitors for what they were about to view in the main chamber, also referred to as the "Blue Grotto." The emotional sculpture of the Virgin Mary, holding the lifeless body of Jesus Christ just after he was taken down from the cross, is perhaps the most famous sculpture in all Christendom. The dramatic robe of Mary flowing to the marble base, which was "the rock of Golgotha," formed a pyramidal grouping. The white Carrara figures stood in sharp contrast to the dark blue velvet drapery that formed the backdrop. Flickering votive lights mounted on vertical metal strips framed the *Pietà* on either side, and 400 lights formed a halo suspended from above. It looked as if the sculpture was floating against the night sky.[2]

To accommodate the thousands of visitors each hour, the exhibit had three moving walks, constructed at three levels, and a fourth stationary walk in the rear. The safety of visitors and the sculpture was of primary concern. Stationary handrails led to the moving walkways. There were foot-level lights and emergency controls for each separate moving walk. A network of protective measures was in place to safeguard the *Pietà*. The most apparent of these was the bulletproof ceiling-to-floor Plexiglas sheets installed to shield the sculpture. Uniformed private guards were hired to keep watch day and night.

On April 19 the *Pietà* was unveiled before several hundred invited guests in an elaborate ceremony led by Paul Cardinal Marella, archpriest of St. Peter's Basilica and papal legate, assisted by Cardinal Spellman. Mayor Robert F. Wagner Jr. and all borough presidents were in attendance, with Lieutenant Governor Malcolm Wilson present on behalf of Governor Nelson Rockefeller. R. Sargent Shriver, then head

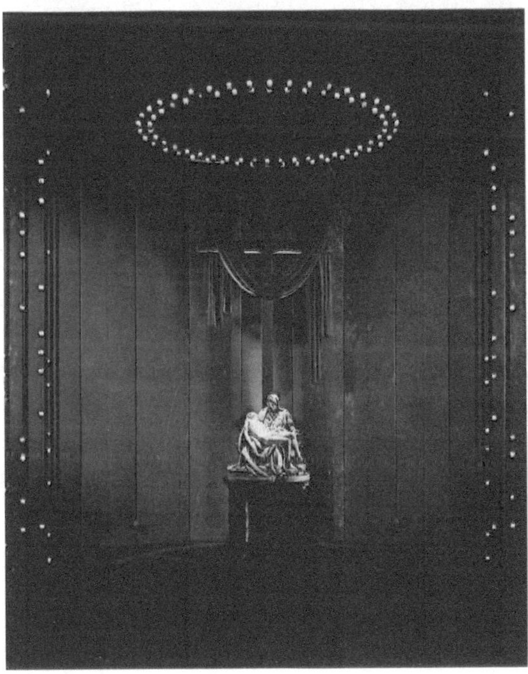

Figure 10.1. Formal photo of Mielziner's staged *Pietà*. (Courtesy of the Diocese of Brooklyn Archives)

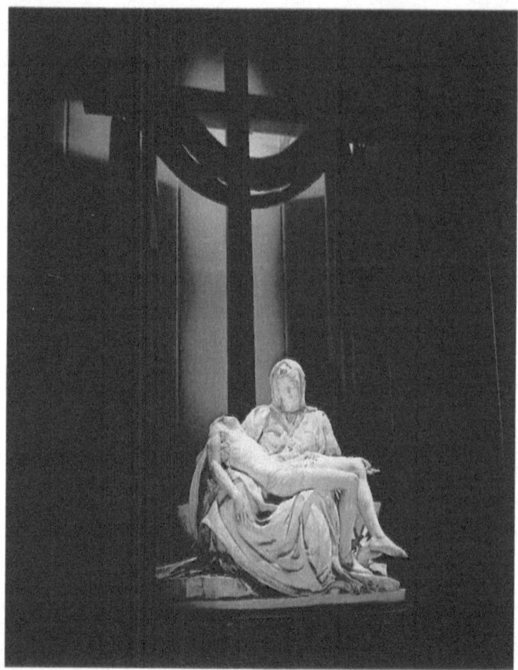

Figure 10.2. Close-up of Mielziner's staged *Pietà* (*Official Guide Book, Vatican Pavilion, New York World's Fair 1964–1965*)

of the Peace Corps, and brother-in-law to the late President Kennedy, represented President Johnson.

Coincidentally, or maybe not, on the same day in Rome, Pope Paul VI marked the four hundredth anniversary of the death of Michelangelo, who died February 18, 1564, with a memorial Mass. Later, the pope, who had a great appreciation for contemporary art, addressed the Sunday afternoon crowd gathered beneath his palace window in St. Peter's Square, saying that "truth and life can be expressed with the voice of our times" by present-day artists, just as Michelangelo had done in the fifteenth century.[3]

More accurately, the New York ceremony was not an unveiling, since the *Pietà* could not be curtained off or draped; it was a "lighting" ceremony. After the pope's message was read, the dimmed lights were slowly turned up brighter, spotlighting the Virgin Mary cradling the corpse of her son, framed by the flickering blue votive lights. Robert Alden, reporting for the *New York Times*, recorded that at first there was "a smattering of applause as the *Pietà* came into view," but that quickly ended when visitors realized it might not be appropriate.[4] Alden also noted that "it was apparent almost immediately that the setting might prove controversial."[5] Part of the problem was the Plexiglas panels installed to protect the sculpture.

After the ceremony, the pavilion was blessed with a procession around the building of the colorfully vested prelates and the ceremonial armed guards, the Knights of Malta. Visitors ascended to the second floor for a Mass in the Good Shepherd Chapel. To conclude the event, Robert Moses hosted a gathering at the fair's Terrace Club. Eloquent as ever, he said, "As the Unisphere symbolizes the world of industry, so the *Pietà* and the other exhibits from the Holy See represent the world of the spirit." Ever ready to admonish his critics, he continued in his now famous quote: "Throughout the fair in all of our trials the critics have kept up their incessant yapping. Critics build nothing."[6]

The critics were only getting started. Among the few hundred in attendance at the *Pietà* unveiling was John Canaday, the leading art critic for the *New York Times*, who made a number of enemies when he lambasted abstract expressionist painters for their "tolerance for incompetence and deception."[7] He was not far off the mark, but when it came to his critique of Jo Mielziner's staging of the *Pietà*, he misfired with his delivery of a scathing review. He wrote that it was "so out of key, so at variance with what the sculpture deserves, that in seeing the *Pietà* here in New York I could feel nothing but a sense of violation." Canaday continued that this complicated work was pulled out of its harmonious surrounding and "placed behind a transparent vacuum." Acknowledging that blue is the color traditionally associated with Mary, he said that here the statue looked as if were "being subjected to refrigeration."[8]

Reading this review, John Walker, the director of the National Gallery of Art in Washington, penned a letter to Cardinal Spellman to express his disappointment in Canaday's review, calling it "ill conceived and unfair." He continued that the criticism ignores the most important consideration, that "the *Pietà* is carrying out the function for which it was created, to inspire with compassion and love millions of human beings." It was *not* made for a museum exhibit.[9]

By contrast, Canaday praised the Asia House's concurrently running display of the *Avalokiteshvara*, the feature of the "Art of Nepal" exhibition. The *Avalokiteshvara* is a bodhisattva who embodies the compassion of all Buddhas. The seven-hundred-year-old bronze statue was shipped from the Golden Temple in Patan, where visitors were not allowed, making this the single chance for New Yorkers to catch a glimpse of it. Like the *Pietà*, it was blessed before it began its own journey, but that was where the similarities ended. In a separate review, Canaday piled on the criticism:

> [The fair degraded] the works of art by making them sideshow exhibits, as we all saw done to Michelangelo's *Pietà* in its famous journey to the Fair. . . . [The negotiations for this bronze were even] more elaborate than those surrounding the extraction of the *Pietà* from its rightful spot in St. Peter's and its mounting, like a huge over-polished tooth in the lamentable setting it now occupies in the Vatican Pavilion. You can get close to the [Avalokiteshvara], walk around it, contemplate it, instead of being carried along a moving belt, slaughter-house fashion, with canned music that makes you readier for the sacrificial knife.[10]

The pavilion directors were later informed that Canaday was disturbed that he had not received the kind of personal attention he had expected. Apparently, in the rush of guests, he had been pushed down a flight of stairs and then provoked a Pinkerton guard, who threatened to punch him in the nose.[11] Mielziner felt it undignified to respond to Canaday, but in a letter to Monsignor Flynn, the pavilion director, he wrote that if the *Pietà* had been shown in a museum, "not ten percent of the daily average attendance could even glimpse it." Would Canaday have preferred "putting the 'Pietà' on a truck and moving it around, so that people would not be 'carried along on a moving belt slaughterhouse fashion' "?[12]

Mielziner knew that displaying a work that had such great spiritual significance would leave some people dissatisfied. He was not interested in presenting the *Pietà* in an art gallery or museum atmosphere. Having worked in the public eye for decades, Mielziner wrote, "I don't think I am over-sensitive, but I have never had in my entire career so much adverse criticism of any work as I have had on this."[13] The time and labor

Mielziner donated to this project were substantial. Overall, the reaction to his setting was appreciated by his peers. The playwright Arthur Miller wrote, "Personally, I see no objection to the blue, starry background nor the evanescent cross."[14] What the critics didn't seem to understand was that the eye-level view, approximately at the shoulder height of the reclining figure of Christ, was probably more satisfying than one could have experienced in St. Peter's at that time.[15]

This was not Mielziner's only contribution to the fair. The Bell Telephone (the parent company of AT&T) hired Mielziner to design its show, which was a fifteen-minute ride that transported the visitor, in a moving armchair fitted with stereo earphones, through the three-thousand-year history of communications "from smoke signals to satellites."[16] Why would Mielziner attract so much criticism from the press for one exhibit, and barely receive notice for the other? The public response to the *Pietà* exhibit, however, was overwhelmingly positive. One foreign correspondent from Paris, Robert Bosc, acknowledged that those Europeans who had already seen the *Pietà* in St. Peter's were impressed. In fact, seeing it in a new setting, it felt like one was seeing the sculpture for the first time. Visitors were moved, spontaneously lowering their voices, even as they left the exhibit and explored the rest of the pavilion. "Such is the extraordinary power of art to transform a holiday crowd, within a matter of seconds, into serious and reflective persons."[17]

Marie McCormack, a former concert artist who served as a recreation director for the American Red Cross, was recruited as the director of the hostess volunteers. She would screen young women, all Catholic, who were more interested in serving the church than making money. The hostesses were chosen from several thousand applicants, with preference for those who were bilingual. As a result, nearly all the hostesses were bilingual, and someone could usually be found who could handle questions in Spanish, or German, French, or even Yiddish. Once selected, they went through training to thoroughly familiarize themselves with the exhibits, especially the *Pietà*. To that end, they also read Irving Stone's recently published *The Agony and the Ecstasy*. The hostesses were dressed in stylish, princess-line dresses of magenta and pink wool mohair, designed by Margarita and José de Lema. To complete their outfits, the young women received a brooch inspired by the papal coat of arms, considered so beautiful that many thought it was it worth the hours of volunteering just to own it. Additional women volunteers were recruited from the Councils of Catholic Women and the Ladies of Charity.

It was the men, however, who were tasked with the critical job of funneling thousands into the pavilion and efficiently managing the wall-to-wall crowds. A *Washington Post* editorial recommended that other exhibitors should study their operations. Under the direction of Christopher Kiernan, who took leave from his job at Pennsylvania Station, more than one hundred men from the Knights of Columbus and the Holy Name Societies contributed thousands of hours of invaluable service. They were responsible for everything from ensuring a smooth flow through the main doors, to counting each visitor using a hand counter, to keeping watch over the potential accident spots, such as the *Pietà* area's moving walks and its ramps, to the pavilion stairways leading to and from the chapel. They also served as ushers at all the Masses and provided assistance to anyone they saw in a wheelchair or on crutches. Thanks to them, in large part, the pavilion could offer special events, such as Catholic Education Day, Catholic Press Day, and Nationality Days such as Ukrainian Day, Polish Union Day, and many more. What was even more remarkable was that these men took time off from their regular employment to commit to fifteen hours a week for twenty-six weeks. This also meant each volunteer covered his own travel, meals, and even parking fees of $1.50 in the visitors' lots.[18]

The pavilion directors soon realized that a car was required for the number of dignitaries from all over the world who paid a visit to the pavilion. As the parking lots created a revenue stream, Moses restricted the use of nonofficial-use vehicles but made a rare exception in allowing Fathers Gorman and Lahey the use of a car on the fairgrounds. The only stipulation was that it be white in color with no identification of any kind. The car was a special courtesy owing to the priests' unique circumstances. For 180 days of the year, for two years, they were virtually captive by day, as at least one of them had to be available during all hours of the fair. They were also on call twenty-four hours a day for emergency sick calls for visitors and working personnel alike.[19]

During the first week, the pavilion welcomed a steady stream of visitors, including celebrities ranging from former president Harry Truman and his family of three to Hollywood actress Arlene Dahl, Ed Sullivan with his family, and Otto von Hapsburg, archduke of Austria. Volunteers answered about two thousand questions a day, gave more than twenty-five thousand individual tours for pavilion guests, including one vice president of the United States, the chief justice and four associate justices of the US Supreme Court, four members of the president's cabinet, and the governors of seventeen states. Foreign visitors included the presidents of fourteen countries, four prime ministers, fifteen foreign ministers, and more than a hundred members of the diplomatic corps.[20] A Mass dedicated to the disabled was offered by Rev. James Foley, the veterans' chaplain, from his wheelchair, to two hundred wheelchair-bound visitors, and special arrangements were made for the blind, who were afforded the opportunity to "see" the *Pietà* by touch.[21]

The music in the Vatican pavilion was provided by the choir of Manhattanville College, which at the time was home to the Pius X School of Liturgical Music. Located in Purchase, New York, the college offered a world-renowned program that taught Gregorian chant to both students and visitors. During the exhibition planning, Cardinal Spellman had contacted the school to provide the pavilion's music. This became the project of Robert Hupka, a recording engineer for RCA Victor and Columbia Records. Hupka volunteered at Manhattanville for over fifty years, overseeing sound recordings for the Pius X School and later the Music School.[22]

Mielziner collaborated with Hupka and the college's Sister Josephine Morgan, director of the Pius X School, to select the works that best suited the exhibit. The first was a medieval *De Profundis*, sung by the monks of the Abbey Saint Pierre de Solesmes, directed by Dom Joseph Gajard and produced by London Records for the album *Gregorian Chants*. Mielziner liked that the a cappella singing was recorded in the abbey itself, giving the music that quality of reverberation often found in towering cathedrals. The second choice was the sixteenth-century composer Josquin des Prez's setting of *Missa Pange lingua*, *Agnus Dei* no. 2.[23]

As the moving walk carried the visitor toward the front of the *Pietà*, the music gradually became more focused, seemingly to emanate from the vicinity of the sculpture itself. This was achieved by placing one speaker above and behind the visitors at each end of the viewing platform. Bozak made the most of the Plexiglas plates that protected the *Pietà* by positioning the speakers so that the sound bounced off the plates precisely at a point in the center so that visitors at either end of the platform would hear the reflected sound from the speaker, while those directly in front of the statue would hear the blended sound of the two speakers, as if the sound came directly from the *Pietà*.[24] As visitors exited from the *Pietà* area and moved into the main chamber of the pavilion, the tone of the music shifted, to lighten the spirit of the moment.

Mielziner worked closely with the architects and the pavilion committee, anticipating problems before they happened. Though the area for which he was responsible was not ready to be turned over to him until a month later than scheduled, he delivered a month in advance. The one problem he did not anticipate was unauthorized tampering with the lighting. One of Mielziner's areas of expertise was stage lighting, and he had labored to develop a highly calibrated lighting master plan. At one point, he discovered that someone had tampered with the entire lighting setup, including eliminating key lights. He asked that either the switches and lighting instruments be locked or that the pavilion committee issue orders to prevent unauthorized experimenting, otherwise he could not continue on the project.[25] Monsignor Gorman responded immediately to assure him that they wanted to maintain his lighting plan and design setup. Gorman was already upset by the electricians supplied by the World's Fair Maintenance Company, with "as many as eight or ten different individuals report[ing] for duty here in a week."[26] In general, the revolving door of workers was annoying, but tinkering with the *Pietà* lighting was most disturbing. Gorman immediately locked the *Pietà* dimmer board and arranged to have a specially trained electrician put on permanent staff who would acquire a thorough knowledge of the lighting scheme and be solely responsible for maintaining the sophisticated lighting plan.

To keep the exhibit looking its best required regular maintenance to adjust the focus and angling of the lamps and to replace bulbs that burned out. After only two months on display, the bulletproof shield, the base, and statue itself needed cleaning. The delicate job of cleaning the *Pietà* required vacuum cleaning with a soft brush by Tommasi Marble Company, subject to approval by the insurance inspectors.[27]

As the first season came to a close, arrangements were made to store and safeguard both the *Pietà* and the *Good Shepherd* over the winter months. The statues would be secured in a specially designed steel cage throughout the fair's off season. Kinney, who was responsible for the care of the statues, had three major concerns. Even though the *Pietà* area of the pavilion was constructed to meet New York City code, fire was considered

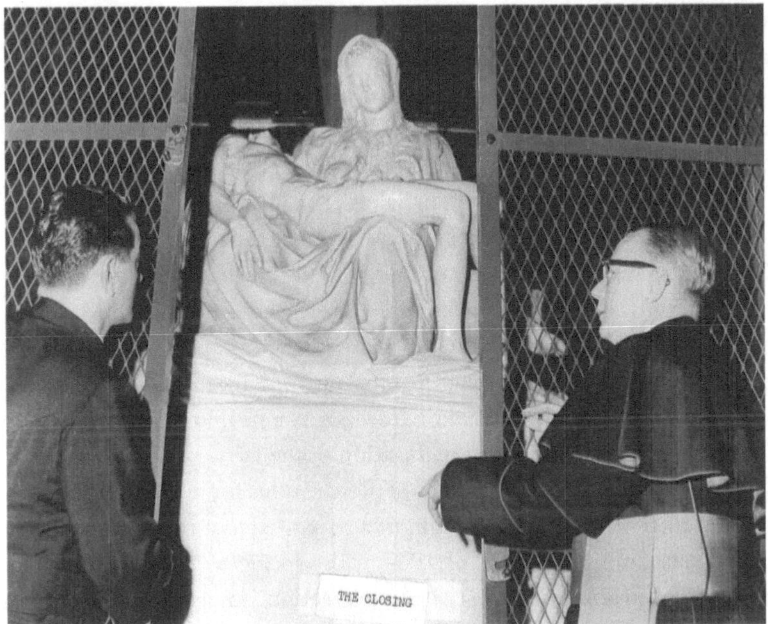

Figure 10.3. Monsignors Gorman (*right*) and Lahey close the protective cage around the priceless *Pietà*. (Photo courtesy of the Diocese of Brooklyn Archives)

the major hazard. To protect the sculptures, Kinney had an asbestos blanket fitted completely around the cage. Also, he installed a separate water line to run from the nearest accessible piping source, with a turn-on valve outside the *Pietà* area. It would function as a sprinkler system that would pour a sheet of water over the asbestos-covered cage if a fire did occur.

The second concern was the temperature change with the onset of winter. Three temperature-controlled space heaters were installed and were set to activate should the temperature within the cage fall below that which the pavilion heating system maintained. The third danger was the possibility that the pedestal on which the *Pietà* sat might shift and add stress to the plaster collar around the thirteen-thousand-pound marble base. To remedy that possibility, two carpenter levels were situated at right angles to each other on the pedestal itself. These levels were routinely read by the security personnel, who stood guard twenty-four hours a day. The cage was connected to an "ultrasonic" alarm system, and two additional alarms were added on the entrance and exit to the *Pietà* area. Only three keys to the enclosure were made, and the holders of the keys were the only persons with authority to shut off the elaborate alarm system.[28]

The *Good Shepherd* was repacked in the inner case in which it was shipped, then placed within the *Pietà*'s protective cage.[29] The security system was described as an "electronic burglar alarm" so sensitive that during a heavy snowstorm, snowflakes that fell through a tiny crack in the door near the *Pietà* and landed on the wires triggered the alarm.[30] Called from his sleep, Monsignor Lahey, associate director of the pavilion, jumped in his car and headed straight over. When he arrived, the area was buzzing with Pinkerton police and New York City detective squads.[31] If the alarms were to ring during the fair season, Lahey would not have far to go, because he and Monsignor Gorman, the pavilion director, took up residence in the pavilion for the duration of the fair.

Here the sculptures wintered until preparations for the fair's second season began. The Brooklyn Museum had requested that the *Pietà* be moved to the museum from the fall of 1964 to the spring of 1965, which would have solved problems of storage and protection. After some deliberation,

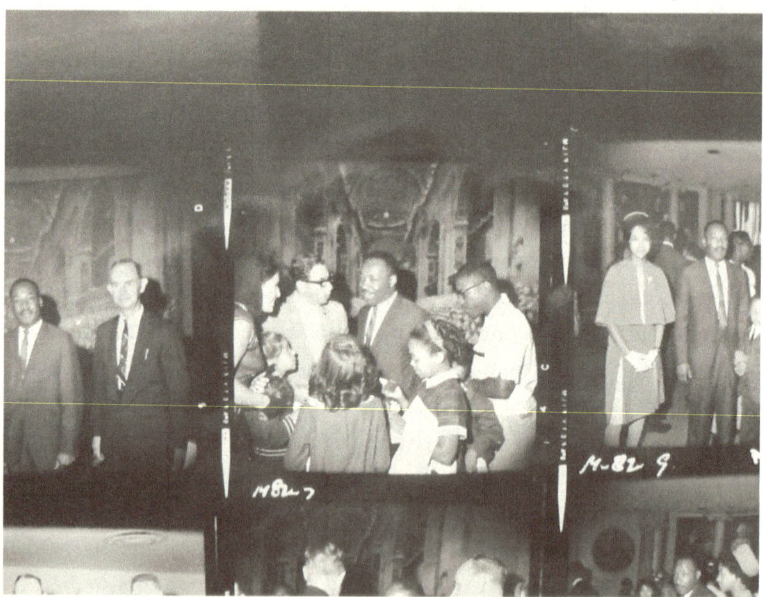

Figure 10.4. Dr. Martin Luther King Jr. is surrounded by well-wishers at the Vatican pavilion. (Photo courtesy of the Diocese of Brooklyn Archives)

however, the directors decided that the *Pietà* should not be made available to viewers during the months in between the two fair seasons. Four or five months in the Brooklyn Museum would seriously reduce the sculpture's drawing power during the second fair season.[32] The *Pietà*'s metal-cage "cocoon," mounted on wheels, was opened and rolled away on April 8. No one, including the insurance underwriters who were present for the unveiling, had much doubt that the statues were unscathed.

When the Vatican pavilion reopened for 1965 it saw a steady stream of visitors, including luminaries such as Dr. Martin Luther King Jr. and Princess Grace of Monaco and her eight-year-old daughter Caroline. On the princess's visit, the *New York Times* reported the key piece of information that she was wearing a "pair of black horn-rimmed glasses," and added that, as she passed the *Pietà*, she was heard saying, "How beautiful!"[33] A photographer caught her and her daughter admiring a new addition to the fair, Pope Paul's papal tiara, a ceremonial crown used during solemn papal functions of a nonliturgical nature. Seven months earlier, at the close of the third session of the Second Vatican Ecumenical Council and before all the bishops assembled in St. Peter's, the pope had made a dramatic gesture: he rose from his throne and placed the tiara on the altar, symbolically presenting it to the poor of the world. The bejeweled tiara was originally given to the pope by the people of Milan on the occasion of his papal coronation—in the days when popes were still crowned. Uniquely modern, it was fashioned in silver and gold, with its three tiers symbolizing the church militant (the church on earth), the church suffering (purgatory), and the church triumphant (those in heaven)—or the teaching, sanctifying, and ruling offices of the pope, depending on which tradition one prefers. One of the themes of the council was relief of poverty, and Paul himself favored the elimination of pomp and display.[34] He then quietly, without publicity, passed the tiara to Cardinal Spellman.

How the tiara, already given to the world's poor, was then passed on to Spellman, no one is quite sure. When Spellman brought the surprisingly light (three-pound) tiara back from Rome, customs appraisers set its value at $10,000, but other estimates ranged from $15,000 to $80,000. For now, it would be added to the Vatican pavilion exhibits, along with a gold filigree stole, a deathbed gift of Pope John XXIII, and candles from Pope Paul as a reminder that Catholics should pray for the success of the Vatican Council.[35] Until the fair's closing, the tiara was on exhibit in the

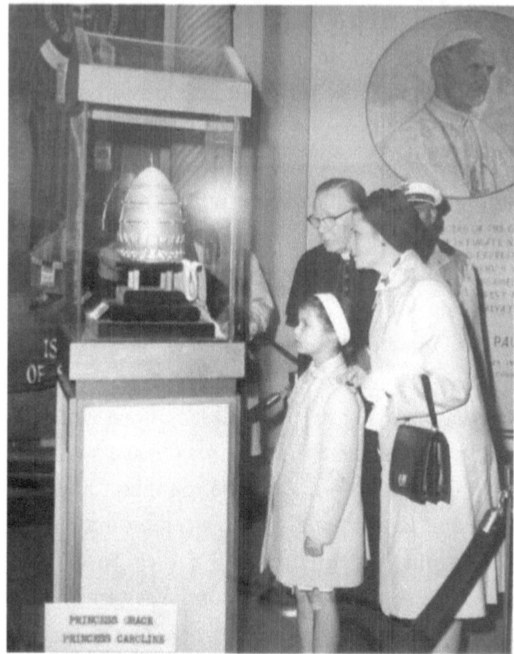

Figure 10.5. Princess Grace of Monaco with her daughter, Caroline, admiring the papal tiara. (Photo courtesy of the Diocese of Brooklyn Archives)

area of Saint Peter's Tomb, under the appropriate theme of "From Peter to Paul." Norman Laliberté was once again enlisted, along with architect Roger Pelliton, to design the display for the tiara's arrival on April 1. Never at a loss for words, Robert Moses thanked Cardinal Spellman:

> These jewels are not merely chunks of carbon and bits of long-imprisoned sunlight and energy fashioned into a crown. They add up to more than science and ingenious craftsmanship, more than dollars to go to charity, more than temporal power and tradition. They are aspects of Divinity. They symbolize the highest aspirations and achievements of man in a changing world. In that sense they symbolize the Fair.[36]

Rather than immediately putting the treasure on sale, Spellman decided to exhibit it across the country under the auspices of the War Relief Services. Donations were expected to surpass its market value.[37]

Adding new attractions to the pavilion would encourage those who visited in the first season to return in the second. An East Side Manhattan art gallery owner, Piero Tozzi, loaned to the pavilion the sculpture *The Young Saint John the Baptist*, believed to be a lost work of Michelangelo. A tender portrayal of Saint John as a boy, the Carrara marble sculpture echoes the mastery of Michelangelo, yet there was no consensus among art critics on its authenticity.[38] John the Baptist was the patron saint of Florence and was often portrayed with a lamb, representing the Lamb of God, a reference to Jesus Christ. In his right hand John holds a shell, anticipating his future role as baptizer of Christ. It was not until 2012, after the piece was acquired by the National Gallery of Art in Washington, that the true author of this work was determined to be Giovanni Francesco Susini, a Florentine sculptor who lived from 1585 to 1653. Susini was trained by his uncle, Antonio, who worked with the sixteenth-century bronze sculptor Giovanni da Bologna, better known as Giambologna.[39]

An added feature to the pavilion for the fair's second season was the Father Mendel Garden of Heredity, in commemoration of the hundredth anniversary of Father Gregor Mendel's discovery of some of the basic laws of heredity. Pavilion organizers landscaped a three-thousand-square-foot garden to demonstrate Mendel's experiments, which were referred to as "one of the great detective stories of all time." Red, white, and blue flowers illustrated "the Case of Elusive Color," tall and short yews were planted to explain "the Case of the Missing Midget," and smooth and wrinkled flowers were applied to "the Case of the Vanishing Wrinkles."[40] To accompany the exhibit, an illustrated handout was available at the pavilion. The garden was made possible through an anonymous donor associated with Fordham University. At the dedication, Moses promoted the garden as "irrefutable evidence that religion and science are not irreconcilable."[41]

Gift Shop

Even though the Archdiocese of New York was responsible for the financing of the pavilion, Cardinal Spellman issued a special collection request to dioceses across the country. Plenty of midwestern bishops must have wondered why they were expected to contribute to an undertaking in New York; nevertheless, the backing was substantial. Supplementing this support were generous donations from pavilion visitors to the chapel mite boxes—but the bulk of revenue came from the pavilion gift shop.

More than 1.2 million customers shopped for gifts and memorabilia, but foot traffic was only one reason the gift shop was so successful. It was the only place one could buy *Pietà* jewelry—in fact, it was the only place one could buy *Pietà* anything. The Fair Corporation gave the Vatican pavilion the exclusive right to merchandize Roman Catholic religious articles at the fair, including the sale of multi-sized *Pietà*s and any articles relating to the Vatican pavilion and its exhibits.[42]

Edward Kinney, who in addition to overseeing the shipment of the *Pietà* was responsible for the gift shop, closely monitored the souvenirs being sold at other pavilions. Whenever he saw a *Pietà* being sold outside the Vatican pavilion, he notified the Fair Corporation, who would issue a cease-and-desist order. Both the Belgian Village and Colonel Morey's Bourbon Street complex were fined for selling these items.[43] The Louisiana pavilion was selling "glow-lite" *Pietà* postcards, and the Better Living Center was selling glass-encased candles bearing a reproduction of the *Pietà*. Kinney offered to buy up the inventory of *Pietà* statuettes that the Artisans of Italy were selling, since presumably they had acquired them in "good faith"—only to turn around and see that they were hawking lithographed pictures of the *Pietà*, so the offer was withdrawn. The image of the *Pietà* was so pervasive that it turned up in the *Official World's Fair Coloring Book*.[44] *Pietà* replicas were a big seller—six hundred thousand in total. They ranged from high-quality plaster casts to glow-in-the dark plastic. View-Master slides, paperweights, ceramic tile trivets, tie clips, night lights, and key chains carried the image. Copyrighted *Pietà* rosaries and *Pietà* medals were offered in two sizes for men and two for women. Other items included pavilion bracelets, silver charms of the *Pietà* and of the pavilion, and even a miniature tiara charm.

Among the souvenirs, the most popular items were the postcards (752,840 sold in total), which, once written and stamped, were collected from pavilion mail drops, flown to Vatican City through a special arrangement with an Italian airline, and franked by the Vatican City post office. In addition to the standard Vatican City postmark, postcards sent in 1965 also received the papal tiara stamp. The gift shop also made available four special-issue Vatican–New York World's Fair stamps. Two stamps depicted Pope Paul VI in three-quarter profile, and the other two commemorated the loan of the *Pietà* to the fair, with one stamp of the Virgin Mary and the second a side-angle view of the sculpture. The shop also sold a series of at least three souvenir stamp sets, specially packaged for stamp collectors. For women attending Mass at the pavilion's chapel, the shop carried mantillas, women's lace head coverings. These were still the years when Catholic women worshipped with their heads covered, even if it was a small circular piece of lace (or a piece of Kleenex).

Since spoon collecting was still a popular hobby, the Norwegian Silver Corporation was selected to provide souvenir spoons for the gift shop; sterling silver or silver plated, they were a popular item.[45] The Italian-born American sculptor Albino Manca designed the official bronze medal commissioned for the pavilion. The medal depicted the *Pietà* on the obverse and images of Pope John XXIII and Paul VI on the reverse. Manca is best known for his bronze eagle sculpture, one of eight standing at the East Coast Memorial in Manhattan's Battery Park.

The shop offered a variety of books, particularly official guidebooks, but also special feature books of the Vatican City in photographs. Kinney was open to the idea of carrying the small hardcover book if the price was attractive;[46] in exchange for the use of the Vatican pavilion name on the book jacket, Doubleday offered a royalty of 1 percent of the book's list price.[47] Irving Stone's *The Story of Michelangelo's Pietà* came out just in time for it to be carried during the second season of the fair. Josef Vincent Lombardo, an art historian at Queens College, also managed to capitalize on the timing with his more scholarly tome, *Michelangelo: The Pietà and Other Masterpieces.*

Even though the fair was operating in the red, the Vatican pavilion gift shop was realizing a profit—a very big profit—enough to finance the visit of Moses's most anticipated guest.

11

City's Embrace

The Pope's Visit

Robert Moses had hoped the pope would visit the fair, but now a papal visit took on a new sense of urgency. The fair was in deep financial trouble, and this was the boost Moses believed it needed. When he and his team first met with Pope John XXIII in 1960 to enlist his support for the fair, Pope John "remarked jokingly, that perhaps a pope may turn up in New York to see the fair,"[1] but he did not live to see the fair open. After Cardinal Montini's election as the next pope in June 1963, Moses's hopes for a papal visit were renewed. Montini had proven himself to be an "intellectual with liberal sympathies," had a passion for politics, and as pope became a Cold War arbiter.[2] The somewhat sudden decision to visit the Holy Land made him the first pontiff to leave Italy in more than a hundred years and, also, the first pope to fly in a plane. This did not go unnoticed by Moses, who issued a press release linking Pope Paul's "precedent-shattering pilgrimage" to highlight the Vatican pavilion's exhibit of the replica of the shrine-tomb of Saint Peter.[3] If the pope would fly to the Holy Land, why not to New York for the fair?

While Spellman was in Rome attending the Ecumenical Council, Moses had enlisted him to secure a commitment from the pope to visit the fair in its second season. If Spellman did approach the pope, it came to naught, but another opportunity arose when Thomas J. Deegan Jr., the fair's executive chairman, made a similar call. Deegan called it a "long, long shot" and asked Moses not to be disappointed if he "[doesn't] score," noting that up until now, "no one else has succeeded either."[4] On December 30, in a private audience arranged by Spellman, Deegan delivered a letter to the pope from Moses, renewing his invitation to the pope to visit the fair. Moses pointed out that in its first season, "14 million persons of all faiths and races" had visited the Vatican pavilion, reflecting not only "a tribute to the beauty and artistic value of the pavilion itself—and the Pieta of Michelangelo in particular—but a spiritual hunger for the eternal amidst all the diversity of modern life."[5]

The *New York Times* reported that after his audience, Deegan was hopeful, adding that "His Holiness seemed receptive." By now, the pope had not only flown to the Holy Land but had also visited India to attend the International Eucharistic Congress. Vatican sources, however, remained skeptical.[6] Less than a week later, the Vatican, through its press office and the newspaper *L'Osservatore Romano*, sought to "cool hopes" of a papal visit to New York City.[7] But "no" to Robert Moses only meant "no for now."

By the end of June 1965, as if in answer to prayer, the United Press International issued a cable datelined Vatican City, reading, "High Vatican sources have reliably reported that due to the urgency of world conditions, His Holiness, Pope Paul VI, may visit New York on October 23rd. He plans to stay only 2 or 3 days before returning to Rome."[8]

Moses's payday had arrived. Although the final season of the fair was to end October 17, Moses was willing to extend the closing date by one week to accommodate the pope's visit. This also meant that most pavilions would be requested to extend their operations. Then, only two months later, the Vatican, through its nuncio, Egidio Vagnozzi, seemingly dashed any hope of the pope visiting the fair in a memo to Moses declining his invitation.[9] However, Vagnozzi did not state that the pope was not coming to New York, just not to the fair. Hardly discouraged, Moses was not going to let this opportunity slip through his hands, especially now that the date for the pope's visit was fixed for October 4, 1965.

The pope's decision to visit New York had its genesis in 1963 when the reigning Pope John XXIII issued the encyclical *Pacem in Terris* (Peace on Earth), his most significant document, which emphasized human dignity and equality among all people. At the time he wrote it, he already knew that he had inoperable cancer. With less than one year to live, he was determined to write a missive on peace, at a time of heightened Cold War tensions.[10] The encyclical rejected the idea of a balance of power and instead called for a new world order, disarmament, the end of colonialism, and a ban on nuclear weapons. It was well received on both sides of the Iron Curtain.

To study the implications of the document, scholars at the Center for the Study of Democratic Institutions in Santa Barbara, California, called for an international conference to be held in 1964. It was at a preliminary planning session for the conference that an adviser to U Thant, the secretary-general of the United Nations, had suggested inviting Pope Paul VI to address the UN's opening conference in New York in February 1965. In the meantime, the pope spoke on this very theme at the International Eucharistic Congress held in Bombay (Mumbai) in December 1964. Pope Paul sent a copy of his address to the secretary-general, and one month later, U Thant formally extended an invitation to him to appear before the General Assembly, which was currently in its nineteenth session. Paul agreed. According to Peter Hebblethwaite's biography of Paul VI, the pope selected the date October 4, the Feast of Saint Francis of Assisi, for its symbolic weight. Saint Francis, referred to as the *poverello*—"the poor one"—was the patron of Italy, of peace and ecology, and beloved for his Christlike nature. In his lifetime Francis had also sought out and conferred with the sultan of Egypt—a possible nod by Paul to the thirty-seven Arab nations at the UN.[11]

This was the first-ever visit of any pope to the Western Hemisphere, or, for that matter, the New World. Then as today, it was an occasion that would bring the city to a standstill and move millions of New Yorkers, regardless of their faiths.[12] However, A. M. Rosenthal, editor for the *New York Times*, wrote that for the paper's staff, October 4 was perhaps "the saddest day of [their] professional lives."[13] A strike was on, and it became increasingly apparent that it would not be resolved in time. The Newspaper Guild, representing editorial and clerical workers at the *Times*, were on strike from September 16, primarily because of a contract dispute over automation. The powerful guild, supported by nine mechanical and production unions, brought the presses to a halt.[14] The visit of the pope was one of the best single stories in New York's history—Rosenthal and his deputy Arthur Gelb wrote up an assignment list, in hopes that the strike would be resolved, but by October 2, two days before the pope's arrival, the gulf between the two sides was too far apart.

The strike was only against newspaper, not book publishers. Bantam Books, one of the largest publishers of paperbacks, saw the opportunity and offered their Midtown offices as a makeshift "city room" for the *Times* reporters. Rosenthal and Gelb asked the reporters who would have covered stories that day to do the same stories for Bantam Books. Bantam hired them as contract writers, who would keep their assignments, not for the *Times*, but for Bantam's "instant book," completely dedicated to the pope's visit and ready for sale by October 7. The Bantam paperback, titled *The Pope's Journey to the United States*, was subtitled *The Historic Record Written and Edited by the New York Times*, and included thirty-two pages of exclusive photographs.

Planning a papal visit overseas was something of a new venture. The Vatican's advance man was Monsignor Paul Marcinkus, the six-foot-four ex-football player and priest from Cicero, Illinois, now a member of the Vatican Secretariat of State. Born to Lithuanian immigrant parents in 1922, Marcinkus, along with three brothers and sister, lived in a duplex, or a two-flat, as Chicagoans call it, in a working-class neighborhood. The Depression hit the family hard, but the parish priest, Father Vaicunas, kept Marcinkus on the straight and narrow. He served as an altar boy and attended Quigley Preparatory Seminary, where he played on the school's basketball team. He graduated from St. Mary of the Lake Seminary in Mundelein, Illinois, and was ordained a priest in 1947. He served at St. Christina Church on Chicago's far South Side for two years, then spent one year in the Chicago chancery office before he went to Rome in 1950 for studies that earned him a doctorate in canon law. In 1952 he was named to the Vatican diplomatic service. During the course of his stay, he became "a fixture" in Rome and a confidant of popes, ultimately serving as the governor of Vatican City. Marcinkus had served successive pontiffs as "a translator, troubleshooter, and bodyguard. . . . He performed those roles so diligently that he was dubbed the 'pope's gorilla' by members of the press," who figured out that by following Marcinkus they could divine which country the pope would visit next.[15]

On his first advance trip to a host country, Marcinkus would explain to civic and church authorities what the pope sought to accomplish. He would use this opportunity to note what the host country hoped for, and then he would return to present a more detailed plan. The Vatican's advance task force would go along the exact route the motorcade was to take, noting any delays caused, for example, by a passing train. It studied the route to afford the pope maximum exposure without, for example, tying up traffic. The task force would know what the pope would do at any certain point, how long he would talk, and whom he would see. A margin of time would be built into the schedule, to allow for delays, but seldom was the pope late for an appointment.

In describing the preparations for a papal visit, Marcinkus said that "the selection of an airline is usually simple: the pope flies the most convenient airline. This has often been the national airline at the country being visited," although in this case, it was Alitalia. "The Pope is a paying passenger just like anyone else." Marcinkus added that "I have most of the original tickets and copies of the others to show that the passenger, Pope Paul VI, is charged the same fare, including taxes, as anyone else."[16] But the pope, of course, wasn't "anyone else." The forward section of the plane in which he traveled would be made into a small salon by taking out a few seats, so that he could receive visitors, read, or just relax, and a bed would be added if the flight were long. The airlines might provide added comfort, but neither the pope nor his representatives would ask for anything special. Accommodations for the press were made at the rear of the plane. In a gesture of airborne diplomacy, Pope Paul sent radio messages to heads of nations he flew over, even "tiny sheikdoms." Though the pope would visit the rear section in a more casual manner, he did not grant interviews. Overall, Marcinkus found one of the most difficult challenges was moving the pope through crowds. "Our experience has shown that the visit of a Pope is unlike that of any other leader," he said. "There is no way to define it, nor is there any way to gauge the emotion of a crowd. Even experienced security men in a host country underestimate this." He added, "I would rather go through the line against the Green Bay Packers for two hours than face the onslaught of priests and seminarians."[17] For the pope's visit to New York, Marcinkus would already be stateside before the pope's plane took flight.

For its part, New York City turned itself inside out to be well prepared. In a review of the accommodations the city made for the papal visit, Sydney H. Schanberg wrote that "it was not a day for keeping books or worrying about money."[18] Millions of dollars were spent, but many hesitated to discuss the cost, "as though it would be sacrilegious," considering who the visitor was. Many services and facilities were donated. A "Papal Visit News Center" was set up in the wing of a high-rise building at United Nations Plaza, courtesy of Alcoa Plaza Associates, and volunteers from two public-relations firms worked the telephones, typed, and mimeographed the latest updates. To ensure that not one car in the pope's motorcade would hit a pothole, the city's Department of Highways assigned a two-thousand-man crew to the streets along the motorcade route one week before the visit. New York streets never looked so good.[19]

Security was the largest cost, as the New York City Police Department assigned more than half its nearly twenty-seven thousand men to the event. All department requests for days off and vacations were canceled, as every single patrolman was needed. Detectives and plainclothesmen were also pressed into service, reported Bernard Weinraub: "Many wearing their uniforms for the first time in years, resembled aging veterans at an American Legion convention."[20] Police marksmen armed with rifles were strategically stationed on rooftops, and windows were ordered shut at key points. Security from other city agencies, such as the Transit Authority and the Port Authority, as well as the United Nations and the State Department, augmented the police force. Five police helicopters were deployed to patrol overhead, and a paramedic's van followed from behind.[21]

To accommodate the additional riders, the Transit Authority added extra subway cars to ease access to Yankee Stadium, where the pope was expected to say Mass, and to other points all along the pope's route. The city's commuter railroads, the New Haven, New York Central, and the Long Island, all followed suit. The pope departed Rome on October 4 at 5:30 a.m., which American television viewers could follow, thanks to Early Bird satellite transmissions, and, crossing time zones, arrived at Kennedy International Airport the same day at 9:30 a.m. In recognition of the pope's special status, the Department of State waived requirements that he present his passport to an immigration officer or open his suitcase for a customs official. There was no protocol for meeting a pope, because a pope had never visited before. Peter Grose, a *Times* reporter, detailed the diplomatic dilemma. The pope is the head of a church, as well as a chief of state, but the United States did not recognize Vatican City. The Department of State wanted to be cautious, as they were wary of establishing a precedent that would later cause difficulties with other visitors; but for the sake of this one-day visit and as a matter of common sense, the pope's status was recognized as chief of state. Furthermore, the pope came to address the General Assembly as a guest of the secretary-general, therefore he was not on an official visit to the United Nations, which was accorded extraterritorial privileges.[22]

U Thant was the first to officially welcome Paul to America. He was followed by Secretary of State Dean Rusk, Governor Nelson Rockefeller, and Mayor Robert Wagner. Five American cardinals were on hand to greet him as well: James Francis McIntyre of Los Angeles, Richard Cushing of Boston, Lawrence J. Shehan of Baltimore, and of course Spellman. After a brief blessing, the pope in a bubbletop limousine and his motorcade departed, traveling at a speed of less than twenty miles per hour so that the waiting crowds, estimated at almost four million, could catch a glimpse him. He chose a longer route, which would loop through Harlem and then downtown via Central Park to his destination of St. Patrick's Cathedral. Moving now down Third Avenue, the motorcade paused in front of the Foundling Hospital, established by the Sisters of Charity in 1868, where the pope waved to a choir of singing nuns.[23]

Continuing south through Spanish Harlem, which had been spruced up for the occasion, the motorcade received especially warm cheers. Turning west onto 125th Street, the little parade entered the heart of Harlem, a detour the pope himself had requested. In an unscheduled stop, the motorcade pulled over at 124th Street and Lenox Avenue, the site of Rice High School, run by the Christian Brothers, to personally hand over a sealed letter in lieu of a visit that the students had requested. The sub-superior of the school (vice principal), Rev. Henry A. Otto, had not known that the pope would stop, saying "we were flabbergasted."[24] The motorcade continued on to Central Park, winding down West End Avenue, where crowds of children were "waving flags and applauding, others standing in awed silence."[25] Already running behind schedule, the motorcade was forced to pick up speed to twenty, then twenty-six miles per hour, too fast for viewers to get a glimpse of the pope.

The man selected to chauffeur the pope's limousine was one Neil J. Smith, forty-eight years old and six feet tall. He was Ford Motor Company's chief chauffeur in New York and was known for his calm and steely presence. This was not his first brush with world figures, as he had already chauffeured Jordan's King Hussein, Prince Bernhard of Belgium, and Jackie Kennedy, but nevertheless he prepared for this assignment by taking several practice runs through the route. Confident, he said, "I can handle just about anything."[26] After the twenty-one-foot-long 1964 Lincoln Continental was test driven forty thousand miles, the custom coachbuilding firm Lehmann-Peterson Inc. of Chicago was commissioned to specially outfit the car, per Vatican specifications, with the removable "bubble" canopy, a public address system, interior fluorescent lighting, and even floodlights. With a newly installed engine, the car could carry thirteen people, including six bodyguards, for whom special running boards and handrails were installed, as well as a three-foot extension to the back of the car.[27]

Finally, the motorcade crossed the park and came down Fifth Avenue, where the sidewalks, as well as the side streets, were jammed with people who had been waiting hours on a cool, forty-six-degree day. A reporter for the *Times*, Martin Arnold, noticed the crowd was on their best behavior: "A neatly dressed man placed his attaché case on the ground beside him. Another man promptly stepped up on it to get a better view. 'Get off it, you—,' the owner started to say. He paused before finishing the word and settled for, 'Anyway, get off it.' A nearby police officer smiled and said, 'You see, it's almost as if they were all in church.' "[28]

Leaving two lanes clear for the motorcade, the police opened up the street to the crowds of people, many of whom had been there since dawn. As the motorcade approached, a tactical force consisting of a hundred specially trained "six-footers" was called in to help the police hold back enthusiastic throngs breaking through the wooden barriers.[29] Arriving at St. Patrick's Cathedral, the pope was welcomed with emotional cheers and applause. Bells tolled, and the organ played "Tu Es Petrus" (Thou Art Peter) as the pope climbed the stairs to the entrance of the neo-Gothic cathedral, where he was met by Cardinal Spellman and other officials.[30]

As the pope walked down the center aisle, the four thousand invited congregants waiting in the pews broke into cheers so loud they almost drowned out the organ's "Papal March." Arriving at the high altar, the pope knelt and offered a prayer of thanksgiving for his safe arrival. Cardinal Spellman then offered a brief greeting, followed by an uproar of approval from the congregation when a priest shouted, "Long Live the Pope!" In attendance were Governor and Mrs. Rockefeller, Mayor and Mrs. Wagner, and the Kennedys: Senator Robert and Ethel Kennedy, and Senator Edward "Ted" Kennedy, with a "radiant" Jacqueline in the front pew.[31] Paul responded to Spellman's greeting and then, in Latin, administered the papal blessing. The pope then proceeded to the north transept door of the cathedral and emerged to crowds on Fifty-First Street shouting "Papa! Papa! Papa!" On their way out, Paul offered his arm to support the aging Spellman, then, waving from the terrace, they proceeded to the cardinal's residence for a private lunch.[32]

New York City was prepared—doubly prepared, not only for the pope, but for President Johnson, who was waiting at the Waldorf Astoria, where

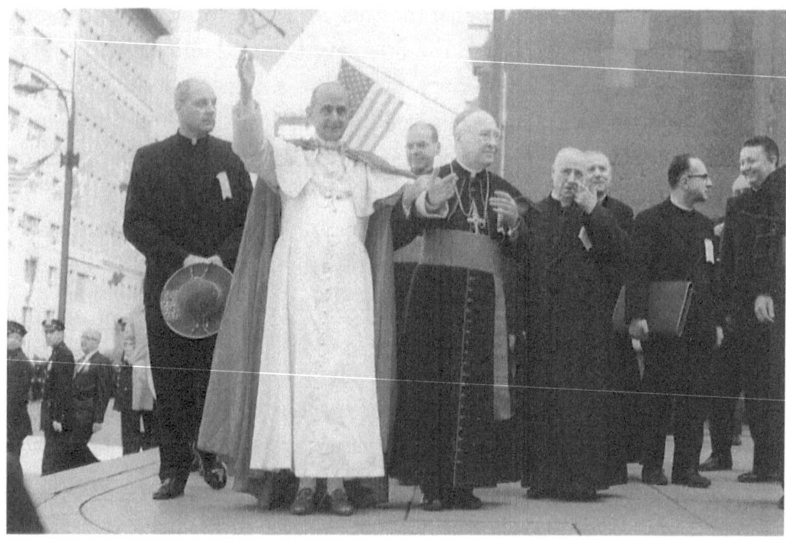

Figure 11.1. The October 4, 1965, papal visit: standing in front of St. Patrick's Cathedral are (*left to right*) Monsignor Paul Marcinkus, Pope Paul VI, and Cardinal Spellman. (Photo courtesy of the Diocese of Brooklyn Archives)

both the New York City Police and Secret Service were on guard. Running fifteen minutes behind schedule, the pope arrived after his lunch and was ushered to his meeting with the president, along with Vice President Humphrey and Secretary Rusk, in the Presidential Suite. The president and the pope then retreated to a small sitting room for a private discussion. Johnson's press secretary, Bill Moyers, was also in the meeting, as well as an interpreter from either side—one from the State Department and the other, Monsignor Marcinkus.

In their forty-seven-minute conference, Johnson and the pope had time to touch on a number of issues, but not enough to cover them deeply. Among those discussed was the India-Pakistan dispute over Kashmir, for which a cease-fire had been declared just weeks earlier through the efforts of the United Nations. Another was the civil war in the Dominican Republic. Fearing the country would become another Cuba, Johnson had sent in the Marines to quell the rebels; he acknowledged the efforts of the papal nuncio in Santo Domingo to secure a cease-fire. They also discussed Johnson's Alliance for Progress, an economic assistance program to Latin America designed to counter emerging communist threats to the region. Johnson, listing his accomplishments, added his concern for promoting worldwide literacy and his support of civil rights. Pope Paul acknowledged the progress the United States was making, and in return Johnson commented on the recent selection of Harold R. Perry as auxiliary bishop of New Orleans, the first African American appointment of the century. Most of these were preplanned talking points, laid out by Johnson's special assistant for national security McGeorge Bundy, to emphasize common ground and minimize division. Bundy also delineated topics not to discuss: diplomatic relations, population control, or any issues under consideration by the Vatican Council or Congress.[33] But the issue that weighed most heavily on the pope, and perhaps on Johnson too, was Vietnam, where the war was escalating.[34] After the overthrow and murder of Ngo Dinh Diem and the assassination of President Kennedy in November 1963, Johnson's decision to send American troops into combat had led to the "Americanization" of the war.[35] On this topic, Bundy emphasized the "great importance to impress the Pope with our passion for peace in Vietnam, and everywhere else. There have been faint indications that not all Vatican circles are persuaded on this point."[36] Roy Palmer Domenico, in his study of the diplomatic relations between the United States and the

Vatican during the Vietnam War era, observed that Paul had a keen interest in the conflict and viewed his role as one of mediator. He could have left it to the Vatican diplomatic corps or relied on informal contacts, but instead he took it upon himself to speak with leaders directly.[37]

Despite any awkwardness, the meeting was described as cordial, and Johnson called it "a stimulating and inspiring conversation."[38] At its conclusion, the president and the pope posed for photographs, and each delivered a brief statement to the press. Johnson, who was reared in the Disciples of Christ Church, always accorded the pope the utmost respect, whether the cameras were rolling or not. To the pope he presented his wife, Claudia—"Lady Bird"—an Episcopalian, and their eighteen-year-old-daughter, Luci Baines, who had just converted to Catholicism.[39] At the conclusion of the visit, the president walked the pope to his waiting limousine, and the pope, waving with both hands outstretched, acknowledged the crush of well-wishers.

The *Times* journalist covering the entire papal-presidential confab was Tom Wicker, the Washington bureau chief, who for the readers' benefit detailed the four times that a president had met a pope. The first was Woodrow Wilson, who met Pope Benedict XV on January 4, 1919, followed by Dwight D. Eisenhower, who met Pope John XXIII on December 6, 1959, both meetings in Rome. John F. Kennedy's audience with Paul VI took place one day after the pope took office in 1963, and then the current visit, the first this side of the Atlantic. After the pope's departure, Johnson immediately returned to Washington, where he watched the pontiff's next event, his address to the United Nations General Assembly, from the White House on television.[40]

To add an extra layer of security on the way to UN Headquarters with its exposed East River access, two police patrol boats and a Coast Guard cutter surveyed the waterway. The city wanted to avoid a repeat of earlier demonstrations, such as that by Cuban exiles against the presence of Ernesto (Che) Guevara, who had addressed the General Assembly on behalf of Fidel Castro the year before. A bazooka shell had been fired toward the United Nations building from the opposite side of the river—falling short of its intended mark.[41]

Dressed in crisp white, Pope Paul VI looked striking against the green marble rostrum of the General Assembly. Before him sat the representatives of 116 of the 117 member nations. The only country to boycott the papal address was (Communist) Albania. As expected, the pope's speech focused on worldwide peace, and indeed, it was the speech of his life. Without directly addressing the conflict raging in Vietnam, the pope declared, "No more war, war never again!"—maintaining a stance of impartiality and disassociating himself from the sentiments of President Johnson and Cardinal Spellman. When dealing publicly with the issue of communism, the pope could afford to play the "good cop," since Spellman was already the bad cop.

He further emphasized the concept of universality, an oblique reference to admitting membership to China, disarmament, and also the development of the United Nations as the "last hope of concord and peace." His emphasis on human dignity—*dignitas humana*—included the controversial matter of artificial birth control, which he dismissed as irrational, cutting down "the number of guests at the banquet of life."[42] Paul distinguished the war on poverty from a war on the poor. This was a disappointment for Catholics hoping for a change, and for Protestants who expected more it signaled that the Vatican would continue to oppose governmental action to control population growth. The pope closed with an appeal for a return to spiritual values to advance civilization. Then came a silent pause, followed by a burst of applause. Even the Soviet Union's Andrei Gromyko was spotted clapping his hands.[43] The *Times*'s Raymond Daniell summed up the pope's presence as an "aura of brotherly love and self-abnegation which for a brief time at least touched and moved his auditors."[44] Pundits reflected on the pope's speech—it would probably not bring any of the world's conflicts to an end but could focus attention on the unity of humankind. For Catholics, they were not called to be "neutralists," but neither should they think that God was an ally for one side or another. It also signaled to American Catholics a respect for their church and their interconnectedness with society at large.[45]

It was a public relations triumph, laying out the church's position on the most pressing issues of the day, but it also served to "steer" the fourth session of the Vatican Council in Rome, scheduled to take up the issue of war and peace the very next day.[46] Immediately after the speech, the pope was presented to five hundred foreign diplomats and guests in the delegates' lounge. Greeting a long line of well-wishers, he spoke in one of three languages—English, French, or Italian. One of the longest exchanges took place between the pope and Gromyko, followed by a deeper discussion in the secretary-general's private reception suite. The other guest who received special attention from the pontiff was Jacqueline Kennedy, who was not included on the original reception invitee list; her name was penciled in pending the secretary-general's express approval, which he gave. There in the suite Paul presented the secretary-general with a Georges Roualt painting, *Christ Crucified*, as well as a diamond-studded cross and ring, which were to be sold with proceeds to support the world campaign against hunger.[47] U Thant, in turn, presented the pope with a United Nations medal to mark the year of International Cooperation, and, as a personal gift, a Burmese silver bowl inscribed with the words "He who conquers himself, he, indeed, is the greatest of conquerors"—an inscription taken from a classic Buddhist scripture, the Dhammapada. The inscription included a commemoration "on the occasion of the visit of His Holiness Pope Paul VI to the United Nations—Maung Thant, Secretary-General, 4 October 1965."[48]

From the United Nations, it was a short jaunt over to the next scheduled event, an ecumenical gathering of nearly a hundred Jewish, Protestant, and Eastern Orthodox leaders at the Church of the Holy Family on Dag Hammarskjold Plaza on East 47th Street between First and Second Avenues. The church, designed by George J. Sale, with its new building completed just six months earlier, was founded to minister to the early Italian immigrant community. The original structure was built on the site of an old brewery stable in 1925, a bit of history that tied in nicely with the church's name. After the current United Nations building was completed in 1952, the church became its parish and required an enlarged presence for that purpose. The parish complex is the residence of the permanent observer of the Holy See to the United Nations and houses the offices of Roman Catholic organizations active at the UN.[49]

Described by *Times* reporter George Dugan as "ultracontemporary," the church's gray granite design anticipated the post–Vatican II spirit of ecumenism favored at the time.[50] Gone was the central crucifix, which was replaced by a commanding sculpture of the risen Christ. Visitors may have been struck by the absence of statuary in the nave or of sacred imagery framing the altar. This lack, however, was more than compensated for by the striking stained-glass windows and ceramic sculptures designed by the Catalan-born artist Jordi Bonet of Montreal. The west wall is dedicated to the Holy Family's flight into Egypt, linking their trials to those of refugees displaced after World War II. Two window panels, the past and the present, lead to the future, represented by the anchor, which is a type of cross and a symbol of hope. Along the same wall Bonet continues the Flight from Egypt narrative in ceramic sculpture. A ceramic mural extends the length of the east wall, depicting the fourteen Stations of the Cross.

John Canaday, the *Times*'s art critic who earlier panned the Vatican pavilion's display of the *Pietà*, took the opportunity of the pope's visit to opine on the state of church architecture. Neither the Holy Family Church nor even St. Patrick's Cathedral escaped his critique. As for St. Patrick's, it was a copycat church—what could possibly be authentically American about church design "frozen" in the thirteenth century? Think what you will, it does resemble what most people consider to be a church building. Not so in the case of Holy Family, whose façade Canaday described as "sleek as any office skyscraper's."[51] The modern façade is streamlined but uniquely patterned with crosses, which perhaps is one of the few clues that it is in fact a church. Despite the abundance of art, "industrial" is the word Canaday used to describe the interior. Canaday tapped into a tension that exists between contemporary and traditional church design, which to this day has not been (nor may ever be) fully resolved. Regardless of its architecture, Holy Family will forever hold a special place of honor in all of Christendom as the first parish church in the entire Western Hemisphere to be visited by a pope.

With the pope's visit scheduled to last only ten minutes, there was just enough time for an exchange of shaloms and gifts. Philip Klutznick, a leader in the Jewish community, presented the pope with an illuminated scroll inscribed with a fitting verse, Isaiah 2:4, "And they shall beat their swords into plowshares, and their spears into pruning hooks; nation shall not lift up sword against nation, neither shall they learn war anymore." At the conclusion of the interfaith meeting, the pope returned to the cardinal's residence to refresh before the main event, Mass at Yankee Stadium.

The use of Yankee Stadium was offered free of charge (usually it commanded a one-day $10,000 rental fee) for the pontifical Mass. The archdiocese covered all the extras. A $25,000 red carpet was rented to surround the specially built two-and-a-half-story altar. Thousands of folding chairs were set up on the stadium grass, and hundreds of ushers, ticket takers, and guards were hired. The police had kept a close watch on the stadium for the past week but swept it for bombs after a tipster reported that he overheard three men talking about planting an explosive. After the second such call received on the day of the pope's visit, October 4, the police were confident no bombs had made it past their one thousand men stationed inside the stadium since the morning before.[52]

As the pope arrived, "the snack bars shut down, the hawkers' voices stilled, cigarettes were put out and women anxiously patted their heads to be sure their hats covered their crowns."[53] By 8:34 p.m., the stadium was now a place of worship, but as the pope entered, standing in his limousine with his hand outstretched, the "church" reverted back to a stadium when the crowd of ninety thousand broke out in cheers of "Viva! Viva!"[54]

For this once-in-a-lifetime event, *Life* magazine was committed to documenting the pope's visit from the "time he got out of bed at the Vatican the morning of October 4 to board his plane in Rome until he landed there again at the end of a marathon one-day visit."[55] The magazine assigned journalist Gerald Moore, along with photographer Bill Eppridge, to cover the Mass. Moore wrote that when the pope appeared at the altar, located over second base, "the stadium exploded in a near storm of flashes from tiny cameras. None of the light would carry far enough to improve a photo, but collectively the flash cameras disrupted the ambient light so much that I heard Eppridge cursing behind me."[56]

All police in the stadium had been given instructions on posture: they had to face the spectators; they were not to remove their hats or take part in any of the ceremonies, by praying or kneeling. They were not to allow themselves to be distracted from their "primary duty of protecting His Holiness."[57] Assisted by seven servers, two "masters of ceremony," and three aides, the pope, wearing white vestments, intoned the Low Mass in Latin but delivered his homily in English. Congregants followed along with booklets containing the responses and four hymns.

In his homily, there were three main points. The first was "You must love peace," and the second was "You must serve the cause of peace." In this spirit, the pope called on all to build up the social order through aiding all those who were vulnerable, the poor, the sick, and the uneducated. And the final point was "Peace must be based on moral and religious principles, which will make it sincere and stable." By contrast, politics cannot sustain peace.[58] Lay lectors read the Prayer of the Faithful in the five official languages of the United Nations. At the conclusion of the Low Mass, the pope blessed a stone, which had been removed from St. Peter's Basilica. This blessed stone was to be placed in the foundations of a new seminary planned by Cardinal Spellman.

At 10:05 p.m.—only twenty minutes behind schedule—with the Mass over, and as the pope drove away, the organ choir and brass ensemble broke out in the "Star-Spangled Banner," signaling the close of the stadium pageantry. Over at the World's Fair, the Vatican pavilion had shut its doors at 4:00 p.m. to allow for a thorough security check.[59] A Catholic high school band was waiting at the main gate when the pope arrived at the fair at 10:25 p.m., this despite Marcinkus's insistence that the pope enter through the so-called Crash Gate, which was only a few yards from the square in front of the Vatican pavilion. Marcinkus had ruled out any possibility of the pope visiting the fairgrounds, and entering through the main gate would be considered a visit to the fair—how crass!—which would diminish the gravity of the pope's presence at the United Nations.[60] Thinking his word was final, Marcinkus clearly did not know with whom he was dealing. Moses was not going to let this heaven-sent gift go to waste. Instead of an under-the-radar visit, the pope was driven through the fair's main entrance, past the Unisphere, and was greeted by fireworks lighting up the sky and bells ringing from the Vatican pavilion. Tens of thousands of people came out that chilly night to catch a glimpse of the pope. Cardinal Spellman was on hand, along with four archbishops, at the Vatican pavilion entrance to receive the pope. Also waiting to welcome him on the red carpet was Robert Moses, flanked by nineteen pavilion hostesses. Moses had rope-a-doped the pope's man.

According to Philip Benjamin, the *Times* reporter assigned to cover this stretch of the visit, the pope, his entourage, and at least fifteen plainclothes detectives were led to the pavilion's main attraction, the *Pietà*, trailed by a TV camera. When he emerged, the pope climbed the staircase to the Chapel of the Good Shepherd, where he knelt at a prie-dieu and blessed those assembled. The pope acknowledged Cardinal Spellman's efforts, which had spared no expense or sacrifice to prepare for this assembly. He even mentioned to Spellman that he thought the *Pietà* looked even better here than it did in the basilica.[61]

As he departed, he waved good-bye from the second-floor balcony to those waiting outside, and as he made his way back down to the main level, people reached out to touch him. His eye caught three children with cerebral palsy, and he paused to rest his hand on them in blessing. Now with his visit completed, he was met with more cheers of "Viva il papa!" The complete record of the pope's public appearances was captured by more than eighty-five television cameramen who followed the pope for fourteen hours. Portions of TV coverage extended throughout the United States, Canada, and Mexico, while Early Bird satellite transmitted coverage to Britain and Western Europe.

With his one-day visit to New York now approaching a close, the pope's last stop was Kennedy International Airport. Security was much tighter on the return trip—the plane had been fueled by inspected tanker trucks, and as soon as the plane's tanks were filled, they were sealed.[62] Baggage was carefully inspected, and the identity of all passengers was confirmed. Reporting for the *Times*, John Cogley noted that priests and members of the press returning to Rome with the pope were required to arrive at the airport hours in advance.[63] Those who had left Rome with the papal party early that morning looked more exhausted than the pope himself, who had been "under the merciless eye of television cameras constantly from the time he set foot on American soil."[64] After the pope boarded and all settled in for the night, the TWA flight took off at 11:30 p.m., only one-half hour behind schedule. Cogley noted that the quiet of the night flight was disturbed when flight attendants served a beef dinner to those who wanted it, then just as suddenly it was time for breakfast. Now that most had been stirred from their sleep, the pope stepped back to the cabin to greet the press and distributed a medal, as well as a souvenir booklet with reproductions of Vatican postage stamps, specially issued to commemorate the historic visit.[65] As he ambled up and down the aisles, passengers congratulated him on a successful mission. To reporters who asked questions, he smiled but said little, a "skill born of years in the Vatican State Department." In fact, in manner and appearance, he had not changed much from the start of the trip, thirty-six hours earlier.[66]

For one day, nothing else had mattered but the pope's visit. He had made no secret of his disapproval of the Cold War and the American involvement in Vietnam, but always maintained that religious convictions, those that inspired Michelangelo's masterpiece, would also bring humankind to seek peace and harmony. This was a message that most mainline Protestants could embrace. John C. Bennett, one of the editors of the liberal Protestant magazine *Christianity and Crisis*, which up to this point looked at the Catholic Church askance, found it troubling that the United Nations and New York City had pulled out all the stops for the visit, and in doing so gave "such prominence to one religious community." He conceded, however, that the Catholic Church had become "to a surprising extent a champion of humanity."[67]

As the last season ended, the fair was deemed a success on all counts, except one—financially, it was in the red. The Vatican pavilion, on the other hand, closed its books with a profit. Thanks to careful administration and the help of dedicated volunteers, on April 1, 1966, Monsignor Asip, the pavilion treasurer, was able to send refund checks to the 136 dioceses that offered financial assistance to the Vatican Committee for the World's Fair.[68] In response, Bishop McEntegart asked, "Do you think some of the Bishops, upon receiving the returned money, will look at the letterhead and take note that it is dated April Fool's Day?"[69]

A final Mass and farewell were offered the last day of the fair, October 17, to all Vatican and World's Fair employees. The pavilion hostesses and employees sang "Auld Lang Syne" from the pavilion balcony. For the thousands of visitors within and outside the pavilion, a tape recording of Pope Paul's blessing was played. The pavilion bells rang from 9:55 p.m. until the doors were closed forever at 10:00 p.m.[70]

The total number of daily visitors at the pavilion was 27,020,857, with 13,823,037 guests in 1964 and 13,197,820 in 1965, making it the second most popular pavilion after General Motors. During the fair season, 2,549 Masses were offered, and 78 concerts and 45 organ recitals were held.

The busiest day of the fair was the final day, when 183,716 visitors entered the pavilion. That works out to an average of 225 guests a minute for the twelve-hour exhibit day. The pavilion gift shop netted $1.6 million. From a survey held in 1965, it was estimated that one-third of the twenty-seven million visitors were people of other faiths.

Even those who did not have the chance to visit the fair were moved by the *Pietà*'s presence in the country. One Mrs. Oakley Childs of Jacksonville,

Texas, unable to attend, wrote to the pavilion to request a large, framed print of the sculpture as a "daily inspiration for [her]." She continued, "I am not a Catholic but grateful you have brought this great inspiring Marble to our Country."[71] It would never happen again. In response to the Italian outrage that Michelangelo's most famous work was missing for the four hundredth anniversary of his death, in September 1965 the Vatican banned shipment of all its artworks.

12

Ciao, New York

The Fair Closes

Before the last visitor left the pavilion and the Pinkerton guards locked the doors, pavilion directors were prepared for the fair's final days' pandemonium. In a letter sent shortly before the close of the fair, Msgr. John J. Gorman thanked the pavilion volunteers for their service and included an alert as the close of the fair approached: "Experience would lead us to believe that several million visitors will attempt to see the fair in the final weeks. Experience also tells us that among these will be varied assortment of souvenir hunters and possible vandalism. Beyond the usual functions of counting and crowd control, every man should be alert and keep a watch out for souvenir hunters and outright vandalism."[1] Gorman asked that fourteen volunteers be on duty at all times: six men at the entrance, two each at the gallery, crypt, and banner areas, and one each in the children's area and the Sistine Chapel replica. He closed the letter by stating, "We have a proud record well-established for the dignity, decorum and courtesy of our Pavilion."[2] Directors of other pavilions were not as well informed. A day after the fair closed, the purchaser of the Wisconsin pavilion arrived just in time to prevent pilferers from loading the pavilion's "Indian Legend" mosaics into their truck.[3]

There remained a tremendous amount of work before the *Pietà* could be returned to St. Peter's and the *Good Shepherd* to the Lateran Museum. Many of the original artworks loaned to the pavilion had to be reclaimed by their owners. The Time-Life transparencies of Michelangelo's Sistine Chapel vault and the *Last Judgment* were shipped back, and the carillon was restored to its generous owner, Schlumerich.[4]

The New York World's Fair Corporation required that participants demolish their pavilion buildings to four feet below grade level within ninety days after the fair closed, so by the summer of 1965, months before the fair closed, Kinney had already arranged for the sale of the pavilion exhibits and displays. Artwork for sale was offered at market value or the highest bid beyond that. Even though the pieces were designed for a Catholic exhibit, much of it had an ecumenical appeal. A rabbi who was building a synagogue in Queens wanted to acquire three of the bas-reliefs on the pavilion exterior, and a Presbyterian minister in Maine sought to purchase Sister Corita Kent's *Beatitudes*, but ultimately the United Church of Christ acquired the latter work for its headquarters in Cleveland.[5] The largest of the Jonynas and Shepherd exterior panels, *The Triad*, found a home at the St. Anthony Franciscan Monastery in Kennebunkport, Maine, where it can be viewed to this day.[6]

Kinney had received offers for the Laliberté liturgical banners, and the highest bidder was a group of Washington, DC, residents, who envisioned a display at the Basilica of the National Shrine of the Immaculate Conception. They, in turn, were approached by a Chicago advertising executive, Earle Ludgin, a trustee of the University of Chicago, who reportedly purchased them for more than $40,000, then donated the set to the university's Rockefeller Memorial Chapel in memory of his late wife. At the request of shrine officials, he did not take possession of them until after the wedding of Luci Baines Johnson, the daughter of President Johnson, who was married in the basilica.[7]

As for the display cases, vending machines, and office furniture and equipment, Kinney sold it all off for half price.[8] Harder to sell was the thirteen-thousand-pound, custom-made marble pedestal brought over from Italy for the *Pietà*, but it did find a home at the Cathedral Seminary

House of Formation in Douglaston, Queens, where it supports one of the two *Pietà* replicas commissioned by Cardinal Spellman in 1965.[9]

No time was wasted in the safe return of the *Pietà* and the *Good Shepherd* to the Vatican, although some hoped the sculpture would continue a round-the-world tour. Moses received at least one request to please send the sculpture back via Brazil, where the state secretary of São Paulo state assured him that this could all be arranged through their own cardinal with no difficulty. Moses politely declined.[10] Only ten hours after the pavilion closed, contractors began the disassembly of the *Pietà*'s display, at eight the next morning. The display's Plexiglas screen and railing were removed first, but the statue would not be packed until later. The *Young Saint John the Baptist* statue was packed up first because it had the shortest distance to travel. The insurance representative inspected the statue for a condition report, then it was packed and loaded into a McNally truck for the trip back to Piero Tozzi's apartment on Park Avenue and 70th Street and inspected once more.[11] Next to be packed, on October 20, was the *Good Shepherd* statue. Instead of using the older, European method by which it was packed for its first transatlantic journey, it was returning home American style, with foam beads and foam boards. The US Plywood Corporation donated its new plastic-coated plywood, "Perma-ply," for the construction of the new containers to replace the originals, which were the worse for wear. By early afternoon the ancient *Good Shepherd* was brought to what remained of the *Pietà* display area, where it was covered with an asbestos blanket, secured, and locked.

The next day, October 21, was spent entirely in preparations for packing the *Pietà*. All packing materials had been readied in the area before the packing was even begun, and insurance representatives were on hand for a thorough inspection. Workers built scaffolding around the *Pietà* pedestal so that it would be level with the truck's tailgate. Then the packing of the inner case began with the bandaging of Mary's damaged hand first, then "slipsticking" the sculpture onto the inner case base. To prevent bead seepage, Dylite foam board was used to line the container, and the base was secured with a collar. Just as before, the sides of the case were secured, and the case was filled with foam beads. A foam top cover was placed over the filled case, and then its lid was secured. During this time, all fire precautions and security systems were operating until the statue was removed from the pavilion.

It was not until Friday, October 29, an entire week later, that preparations were ready for the inner wooden case to be loaded into the steel container. The pavilion was a temporary structure, but the demolition could not begin until the *Pietà* had left the building. Kinney noted that it was "with little compunction" that the workers tore open a hole in the pavilion wall to allow a McNally flatbed truck to back up directly into the building, as close as possible to the scaffolding on which the steel container rested.

Instead of using traditional rollers, Frank McNally selected four "Rol-A-Lifts," hydraulically powered hand trucks, to move the wooden crate into the steel container already in the truck. Two of the hand trucks were on either end of the wooden case, serving as casters, and a heavy cable placed around the case was pulled by a tow motor, carefully easing the case into the open-ended steel container. Once the wooden case was in place, and cushioned by eight inches of Dylite foam panels, the steel container was closed and locked shut. After the *Good Shepherd* was carried to the truck, both containers were lashed tight. The gaping hole in the pavilion wall was covered with a canvas tarpaulin, with the flatbed still parked inside. Security guards manned the wall opening around the clock Saturday and Sunday, standing watch before moving day on Monday, November 1.[12]

The *Challenger*, the same derrick barge that carried the precious cargo on its original journey, was positioned and waiting at Flushing Creek for the McNally truck to arrive. It was a chilly day, but at least the crew did not have to contend with rain. At 12:30 p.m. the McNally truck pulled out of the pavilion. It left the fairgrounds and made its way up to the Parkway Bridge over Flushing Creek, with members of the transport committee and a police escort at its front and back. The crew unfastened the *Good Shepherd* first, and it was the first to be lifted by net onto the *Challenger*'s deck. When the slings for the *Pietà* were arranged to the satisfaction of Kearns and the barge's captain, the steel container was lifted up over the bridge railing and down to the deck.[13]

River traffic was halted until the barge lift was completed. McAllister tugboats towed the *Challenger* down the East River and through Hell Gate, the narrow tidal strait that separates Queens from Randalls and

Wards Islands, at high slack tide, when the river flow was calmer. From there the route was around Battery Park at the tip of Manhattan and up the Hudson River to Pier 90, where the crew secured the barge overnight for the arrival of the *Cristoforo Colombo* early the next morning.

Kinney knew that on the westbound voyage from Italy, every day had brought the *Colombo* closer to better weather. Now, heading east, the reverse was true. Crossing the Atlantic in November was especially daunting, so extra precautions were taken. These included supplementary stand-by cables fastened to the hull of the ship in the event of an emergency.[14] Once the *Colombo* arrived, the *Good Shepherd* container and then the *Pietà* were secured onto the cabin-class deck. A signaling beacon was placed in its cradle on the top of the *Pietà*'s container.

With the cargo locked in place, the team was able to join a cocktail party in the ship's lounge. This time the wives of the team members were able to enjoy the trip alongside their husbands. Before the Atlantic crossing, the *Cristoforo Colombo* had an eight-hour layover in Boston. During

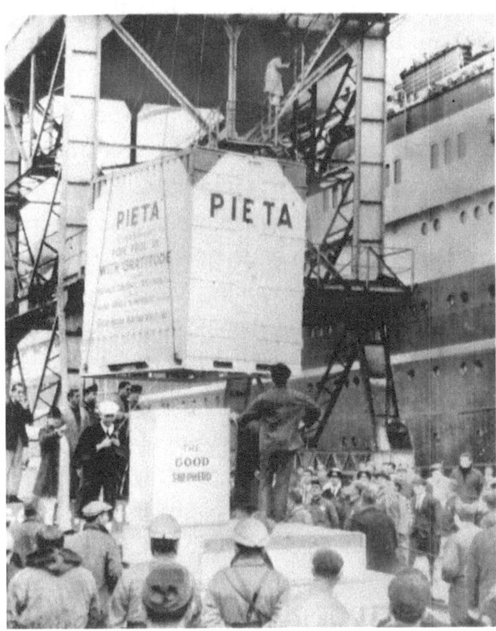

Figure 12.1. Containers of the *Pietà* and the *Good Shepherd* being loaded for the return trip. (AP wire photo)

the entire voyage, the cargo proved to be as big an attraction on water as it was on land. Passengers, both individually and in groups, had their pictures taken standing in front of the container. To accommodate them, team members removed the canvas cover to reveal the white steel container with the letters P I E T À emblazoned across it.[15]

The rough transatlantic weather the team had anticipated never materialized. After days of beautiful skies and calm seas, the *Colombo* made port in Lisbon, where Kinney left the group to fly directly to Rome to ensure that all arrangements were in order on the receiving end. Kinney had written to Vacchini with their expected arrival date but had not received a reply and did not want to risk any surprises. Vacchini, ever reliable, was ready. In Rome, Kinney, joined by Kearns, took the opportunity to make a round of courtesy calls to Count Galeazzi and all others who had facilitated the safe transport.

The *Colombo* was to return to the port of Naples, from where the Gondrand truckers would be ready to transport the statues back to Rome. Poor weather accompanied the ship as it arrived in Naples at 8:00 a.m., but it did not slow down the unloading. The entire unloading was completed by 10:30 a.m. By noon, the convoy left with a dozen motorcycle policemen to accompany it from Naples to Vatican City. The only difference on the return trip was the use of a flatbed truck with a longer wheelbase, similar to the one used by McNally in New York. A longer, lower rig meant a lower center of gravity and more stability and would also reduce the drop if a catastrophe were to occur.[16] As the little convoy made it to the Naples–Rome autostrada, they were met with a downpour of rain. They arrived in Rome by 8:00 p.m., where city drivers showed their dissatisfaction with the slow speed at which they were driving by angry shouts and honking horns. Didn't the Romans know what was being brought back? As the convoy approached Vatican City, its first stop was at the nearby railroad station, where the containers were given a cursory inspection. Once that formality was completed, the truck and its precious cargo spent the night in a Vatican City courtyard, under guard.[17]

A morning session of the Second Vatican Council was scheduled the next day, which meant about two thousand people filling St. Peter's Square, so the delivery of the *Pietà* had to wait. A smaller truck brought the *Good Shepherd*, and Kinney and his team carried the container "sedan-chair style," with two carriers in the front and two in the back. On their way to the Vatican Museum, Kinney noticed that the screws holding one of the carrier handles had loosened, and he was afraid that if the handle came off on one of the sides, the container would hit the ground. Although the drop would have been only about four inches, and the packaging could easily withstand it, it would have caused plenty of embarrassment. They set the statue down on the gravel path, tightened the screws, and, in an abundance of caution, wrapped the entire box with rope. At noon, the flatbed truck brought the *Pietà* to the entrance of St. Peter's. Vacchini had wooden planks installed as a ramp up to the portico to allow the truck to pull in close to the entrance. Cardinals Spellman and Marella formed a welcoming committee at the front door. Others had assembled to watch as well. The truck driver pulled up to the planks and carefully centered the wheels, but because of the intermittent morning rain following the continual rain of the night before, halfway up the first set of planks—to everyone's horror—the truck began to slide back down. Kinney despaired, "We had brought both statues thousands of miles safely. We were just a few hundred feet from the end of the journey, and now this!"[18] The *Pietà* could come crashing to the ground if the truck slid off the planks.

After sliding about eight feet, the truck came to a stop. The driver tried to start it again, but "it shuddered and started to skid from side to side."[19] Kinney knew what had to be done—they needed sand. It only took a few minutes for a workman to bring a wheelbarrow full of dry sand, which was spread on the planks and the wheels. The driver's supervisor gave him the signal to start again, and without any instructions to do so, the transport team members, Gondrand's men, and at least twenty onlookers lined up on either side of the truck, "put their shoulders to it," and supported the truck as it continued up the planks. It made a great photo for the European press. By this time, the small group of onlookers waiting at the portico, including Count Galeazzi, had grown, and now offered their congratulations to the transport team. But Kinney knew their job wouldn't be done until the sculpture was securely resting back on its pedestal. The flatbed had enough room to back up to the scaffolding that Minguzzi had custom-made to the truck bed's height.

The team opened the steel container and removed the Dylite foam panels. McNally used the hydraulically powered hand trucks to raise the wooden case about two inches from the floor, a cable was looped around it, and a portable winch pulled the inner box onto the scaffold. Now, under Vacchini's supervision, Minguzzi and his men took over. A *carrello* moved the case to the platform abutting the pedestal. Before the final unpacking, the team took a one-hour break, then returned to the chapel and began disassembling the crate. Kinney observed the look of "incredulity" on the faces of Minguzzi's men as the marble form began to appear bit by bit as the team carefully vacuumed away the protective foam beads.

When the *Pietà* was finally positioned on its original pedestal, it was 8:00 p.m. It was now safe for Kinney to unwrap the bandages and splints protecting Mary's hand, the sculpture's most vulnerable part. After thorough inspections from the insurance representative and Vacchini, it was apparent that the *Pietà* had made the nine-thousand-mile round trip without so much as a scratch on its surface. *Deo gratias*. Each member of the team felt great relief and satisfaction that the job was completed, and perhaps even a little sadness now that the challenge and excitement were over.[20]

The next day was Sunday, and a day to rest. The pope was to personally thank the team, and for such an occasion the men wore dark suits and the women wore black dresses and mantillas. Just before 5:00 p.m., when the pope would meet them, Cardinal Koenig of Vienna was waiting with the Vienna Boys' Choir, which broke into song. Then the pope, accompanied by Cardinals Spellman and Marella, arrived at the chapel. The pope thanked each member of the transport team individually. They were the first people to ever move the *Pietà* from Rome, and they would probably be the last.[21]

That evening, Cardinal Spellman had the opportunity to show his own thanks by hosting a dinner for the team members, their wives, Brooklyn's bishop McEntegart, who worked so diligently on the pavilion, and the Vatican staff members Galeazzi and Vacchini, as well as other church officials who assisted with the process. Following the dinner were speeches of gratitude and acknowledgment. Kinney gave special thanks to Vacchini, who once he finally had accepted that the Americans were in charge of the move assisted in every way that he could. Then the Americans started a singalong, which the Italians enthusiastically joined in.

Years later, Kinney, when asked if he would ever take on such an assignment again, hesitated. He was older and maybe not quite as adventurous. But the point was moot, because in September 1965 the Vatican approved strict regulations against the loan of its works of art to anybody, anywhere. There was some wiggle room for artworks donated to the Vatican, if the loan was for a religious cause. Those requests for purely cultural or artistic motives, however, or "other purely worldly motives," would not be considered. The *New York Times* speculated that these new regulations were driven by those who had opposed the shipment of the *Pietà* in the first place, and suggested that Pope Paul VI agreed. Furthermore, the new regulations stipulated that any art that would be shipped would be handled by the "Direction General of Pontifical Monuments, Museums and Galleries."[22] It was not until the pontificate of John Paul II, more than a decade later, that a single piece of art left the Vatican Museums or any other Vatican institution.[23]

Back in Queens, the team was busy clearing the site. Several requests were made for the building itself. Some hoped the structure could be salvaged, perhaps for use in mission territory where Catholics were few and funds were not available to build a church. The concrete-and-steel structure, however, could not be taken down in sections for reassembly—but what could be saved was saved, if only in spirit. The Good Shepherd Chapel was reincarnated in Groton, Connecticut, at the newly formed St. Mary, Mother of the Redeemer Catholic Church. The architect, William F. Herman Jr. of Mystic, Connecticut, specifically designed the church to replicate the original chapel. The church stands 172 feet above sea level, and the circular nave allows a panoramic view of the scenic surroundings. The pastor of the fortunate parish was Father James A. Curry, who was struck by the windows when he visited the pavilion on a fall day in 1965. At that time, he was occupied with the architectural details of the proposed Groton church and was struggling with a plan for the windows. In an interview with John Foley of the *Catholic Transcript*, Curry said, "We were looking for glass that would deliver the proper effect, the effect that we had to have, but we just couldn't seem to find it. When I saw the light pouring through those windows, I knew I had found what I wanted."

Why stop at the windows? During his visit to the fair, Father Curry had seen so many other splendid features, including the famous golden cross that topped the pavilion. He wondered what would happen to these one-of-a-kind pieces after the fair closed, and if there was a chance that some of them could be purchased for the new church. After all, he recalled, "the chapel, when I saw it, was very similar to what we had in mind, from a design point of view." Curry took his plan to the pavilion administrator, Christopher Kiernan, who agreed to it. Not long after, the Diocese of Norwich acquired the items, which were then shipped to Groton.[24] The stained-glass windows, from the close of the fair in October 1965 until their installation, were stored in a warehouse a few miles from the Groton church. The same glaziers who installed the windows in the Vatican pavilion placed them in the church. Except for two panels on either side of the altar, all of Jean-Jacques Duval's windows were reinstalled and covered most of the outer wall. There are two sections to each window, an upper and lower register, representing Christ the King of the Universe and the Stations of the Cross respectively, about sixteen feet across, and reaching from the ground to the fluted roof. The windows look even more magnificent in their new setting, since at the pavilion the lower register had been covered by drapery. When opened up, as they were intended to be, they give off a greater sense of light and transcendence. Fifteen newly made, boomerang-shaped laminated wood arches were installed, separating the window panels and supporting the roof.[25]

The original walnut pews were also purchased for the church, as well as the striking reredos, the fan-shaped spray of walnut rising behind the altar, resembling a baldachino, and the halo-like testor, with recessed lights, to illuminate the altar.[26]

What Groton missed was the 1964 M. P. Möller organ. This unified organ had eight ranks and was from the firm's popular Artiste series. When the fair closed, the organ was moved to the Immaculate Convent of the Sisters of Saint Francis in Hastings, New York—the sisters got to it first.

St. Mary's groundbreaking took place on December 11, 1966, and the first Mass was celebrated in the completed church on March 24, 1968. At the dedication six days later, Bishop Vincent Hines described the church as a "beacon of faith."[27] He may have been inspired by Cardinal Marella, the legate of Pope Paul VI, who presided at the unveiling ceremony of the *Pietà* at the Vatican pavilion and had referred to the lantern cross that crowned the pavilion as a beacon that would shine over the World's Fair—"a beacon light of peace and not of battle—a beacon that will shine with the light of Christ, who is the *lux mundi, lux veritatis*. We pray that all may see it."[28] And we still can. The local paper observed that the pavilion cross on the contemporary round church was already attracting visitors and fast becoming a landmark.[29]

Now that the fair had ended, Moses wanted to mark where the Vatican pavilion had stood and wrote to Monsignor Gorman on May 9, 1966, to suggest that a "relatively small dignified marker or memento" be considered for the spot. Moses had already decided on a street sign, but wanted something more noticeable and engaging, such as a monument with an inscription and a symbol.[30]

The fairgrounds were quickly evolving into the park Moses had originally envisioned in the 1940s. The site plan, essentially unchanged since 1939, remained, with fountains, sculptures, and utilities, but the millions of dollars that had been poured into the site for the latest fair resulted in major improvements, such as roads and buildings. Less than two years after the close of the fair, Moses officially turned over the fairgrounds to the city on June 3, 1967, in a dedication ceremony of the new park, renamed Flushing Meadows–Corona Park. Moses's dream to turn this onetime garbage dump into a public park had been realized at last.

The dedication was held at the Singer Bowl Arena. Seated on the dais were Thomas J. Deegan Jr., chairman of the Executive Committee of the World's Fair; Queens borough president Mario J. Cariello; Parks Commissioner August Heckscher; and Lieutenant Governor Malcolm Wilson. Mayor John Lindsay was unable to attend because of an ongoing strike, but his wife, Marion, came in his stead. The Department of Sanitation Band provided the music. After the "Star-Spangled Banner" was played, heads were bowed for Monsignor Gorman's invocation.

Although Moses presided over the ceremony, he did not make a speech but introduced the guests who did. Heckscher, who proved to be no less verbose than Moses, said, "Mr. Moses, I notice, has called this park the fairest garden to be created since the Garden of Eden. If this is true—and who am I to dispute Mr. Moses, my great predecessor as parks commissioner. Then, if it is true, Mr. Moses has led us right into the promised land. Thus being able to do better than his great namesake who could lead people to only within a distant view."[31] After the speakers had finished, Moses lowered the flag of the World's Fair Corporation and raised the New York City flag. At the conclusion of the ceremony, Rabbi Malcolm H. Stern of the Central Conference of American Rabbis delivered the blessing, and a Methodist pastor, Ralph W. Sockman, delivered the benediction.

This day also marked an expansion of attractions to the park, including football and soccer fields, baseball and softball diamonds, a zoo, a children's farm, a horseshoe pit, an ice- and roller-skating rink, a swimming pool, basketball courts, an eight-hole putt-putt course, and a day camp for the mentally disabled, a first in the city. Moses was besieged by autograph seekers and in an interview said, "I feel happy that this has been carried through. I have the same emotion you always have when something you have been working on for a long time comes to fruition." He acknowledged that it was always his goal to repurpose the fair's 646 acres into the "nucleus" of the new park.[32]

One more dedication was scheduled following the Singer Bowl ceremony. Near the southeast corner of park, a monument was erected to mark the site of the Vatican pavilion and the visit of Michelangelo's *Pietà*, and to commemorate the visit of Pope Paul VI. The dedication took place almost two years to the day after Paul visited the fair on October 4, 1965. The elegant monument was in the form of an exedra, a semicircular granite bench, positioned on a stepped, circular platform. The exedra was first

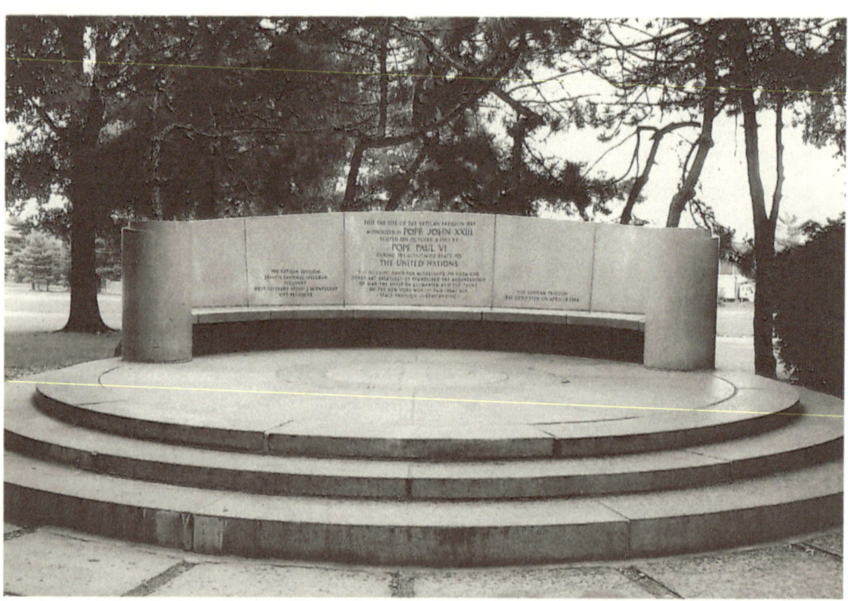

Figure 12.2. The exedra in Flushing Meadows–Corona Park marks the site of the Vatican pavilion and commemorates the pope's visit. (Photo by the author)

designed by ancient Greeks for areas of worship and as a contemplative refuge for poets and scholars.

The cost of the exedra was defrayed through private contributions. At the dedication, Moses addressed his gratitude to Cardinal Spellman:

> We owe this most notable exhibit at our Fair to your generous guidance and indefatigable support. If you had not spoken for us at the Holy See, we should not be here today to memorialize a great occasion and the opportunity vouchsafed to people of all creeds to see, and not as strangers, what is generally acknowledged to be the most moving sculpture in the civilized world. If there were any who came to scoff, they remained to pray. In the absence of an official government exhibit, we had the generous and impressive blessing of Pope John XXIII and the unforgettable visit of his successor, Pope Paul VI.[33]

However unlikely, the exedra has become a pilgrimage site and has attracted as many as fifteen thousand faithful, while vigils are still held on feast days throughout the year.[34] Whether one is a member of the faithful or not, a visit to Flushing Meadows–Corona Park is more than a day at the park—it is a trip back to a magical time when not only millions of people from around the world came to Flushing, Queens, but also Michelangelo's *Pietà*, Our Lady of the Fair.

Epilogue

End of an Era

The end of the fair was the closing of an era and the last project Moses led, his last hurrah. Remarkably, the fair did not produce a profit, and no satisfactory explanation has ever been offered; but it became apparent after the close of the first season that there was a serious accounting lapse. For this and other perceived faults, the *New York Times*, which did not especially like Robert Moses anyway, refused to reward him with praise for the fair, despite its immense popularity. Author and onetime reporter for the *Times* Gay Talese got the measure of the relationship in this passage from his best-seller, *The Kingdom and the Power*:

> Moses, the symbol of the New York Fair, had made too many enemies during his long career. He had written too many letters, pushed too many people. And he got what he deserved, even though, as is often the case, he did not get it when he deserved it. For the New York World's Fair of 1964–65 was not really the ugly, dull, uninspired extravaganza that much of the press coverage indicated. Each day thousands of visitors greatly

> enjoyed the Fair, found the sights and sounds both marvelous and memorable, but they had no way of expressing this, no voice that could compete with a press that focused on the demonstrators at the gates, the problems of parking, the labor disputes, the flaws that can always be found if one looks for them.[1]

For decades, both Spellman and Moses had received privileged treatment from the *Times* but had gradually fallen from grace. Just as the new guard at the *Times* pulled their support for Moses and his authoritarian leadership, they might have continued the "polite press" for Spellman, but his 1966 Christmas comment to American troops in Vietnam that anything "less than victory is inconceivable" came at a time when Pope Paul was working toward a negotiated peace.[2] In response, the Vatican tempered its support for Spellman, while younger bishops were emerging more in tune with the spirit of the council. To quote Russell Shaw, "It became increasingly clear that Francis Spellman's day was passing," though as he approached death, Spellman had mellowed, appealing for racial equality and ecumenical outreach to the Jews.[3]

When Spellman died on December 2, 1967, the front-page news was reported around the world. The Vatican daily, *L'Osservatore Romano*, wrote that Spellman was "one of the most eminent figures among the ecclesiastical dignitaries of our epoch," while the communist press referred to him as "one of the most eager advocates of the aggressive policy of the most reactionary circles of the United States and an active promoter of anti-communist agitation."[4]

The historian Thomas J. Shelley wrote with some perspective that Spellman was an ambitious man, but he directed that ambition for the good of both the church and the country.[5] For those who lived in New York and were educated in Catholic schools, healed in Catholic hospitals, and cared for in Catholic homes for the aged, many would agree.

With Moses no longer in charge of the city's parks, the newly elected mayor, John Lindsay, focused his efforts elsewhere, and it was not long before the Flushing Meadows–Corona Park, abandoned to vandals and the elements, began to suffer.[6] The fountains stopped working and became overgrown with weeds. Fixtures rusted, while structures meant to stand were demolished. Compared to Central Park in Manhattan, the Flushing Meadows–Corona Park, serving the working-class community in Queens, became, as Michael Powell noted, the downstairs of the "Upstairs/Downstairs" city park system.[7]

Decades after Moses died in 1981, his vision for the park has finally been fully realized. While no one has yet found a way to repurpose the Philip Johnson–designed New York State pavilion, it was added to the National Register of Historic Places in 2009. The Federal Building made way for the Arthur Ashe Tennis Stadium, home of the US Open. The Theaterama became the Queens Theater, and the original New York City Building built for the 1939 fair is today the Queens Museum. Moses got the last word, as his signature monument to the fair, the Unisphere, still stands.

While the *Pietà* returned to St. Peter's no worse for the wear, on May 21, 1971, an emotionally unstable Hungarian, Laszlo Toth, thirty-three years old, believing himself to be Jesus Christ, attacked the sculpture with a hammer, breaking off the left forearm of the Virgin Mary, which fell to the floor and broke into pieces. He continued to swing at her head, chipping off parts of the nose, her left eye, and veil. In assessing the damage, the president of the Vatican commission doubted that they could repair the eye and eyelid with accuracy. However, restorers could reference an exact replica of the *Pietà* made before it was shipped to New York, which was displayed in place of the original. Recalling the "painstaking security precautions" implemented to ship the *Pietà* to New York, *Il Messaggero* of Rome criticized the Vatican for inadequate security—a belated acknowledgment of John Kinney and his team.[8]

The *Pietà* was exposed to yet another potential peril, as told by the international art dealer Daniel Wildenstein in his book *Marchands d'Art* (2002).[9] Families had often bequeathed their artworks to the Vatican, and the Vatican in turn would call on Wildenstein's services to sell the works, with proceeds going to Caritas International, the charity arm of the Catholic Church. On one of Wildenstein's visits to the Vatican, his host, Monsignor Rodhain, the head of Caritas, told him that "the Holy Father would like to see you." "I'm honored," he replied. "But does he know I'm Jewish?" Rodhain replied that of course he did, but what of it? Rodhain accompanied him to the pope's private apartment, where the pope was waiting. Paul VI thanked him for coming and asked for his help. Wildenstein received the shock of his life when he heard Paul explain that out of his concern for poverty in the world, he wanted to sell the *Pietà*.

Although the Vatican roundly denied the account, Wildenstein had met with Pope Paul twice, just three weeks before Paul's death on August 6, 1978.[10] The idea of selling the *Pietà* was shocking, but it was not inconsistent with Paul's deep concern for the poor. The pope had already displayed this impulse by his gift of the papal tiara to Cardinal Spellman. If he understood that he was close to death, he may have felt less constrained to make such a grand gesture. If the *Pietà* were sold to a museum, it would still be accessible to viewers but removed from its devotional context and original environment. Had the sculpture been purchased for a private collection, visitors from around the world might be forever denied.

Moses's fair was the last of the grand-tradition expositions in the US—none of the fairs to follow would compare in size or scope. The internet, bringing the world to our living rooms, replaced much of the fair's raison d'être, but technology could never replace the experience. One could practically rub elbows with world leaders and celebrities. Eva Holzapfel of Huntington Station recalls the shah of Iran passing in a golf cart, waving to those waiting to see Eva's favorite exhibit. Likewise, for all those who could not afford a trip to Rome, the fair brought the *Pietà* to them. Only through the extraordinary cooperation between two of New York's most powerful leaders, Our Lady of the Fair made her appearance on a former dumping ground, transformed into a showcase of American optimism and a refuge for an ever-increasingly anxious world.

Notes

Preface

1. Alfred Heller, *World's Fairs and the End of Progress: An Insider's View* (Corte Madera, CA: World's Fair, 1999), 11.

Introduction

1. Lawrence Samuel, *The End of the Innocence: The 1964–1965 New York World's Fair* (Syracuse, NY: Syracuse University Press, 2010), xvi.

2. Hilary Ballon, "Robert Moses and Urban Renewal: The Title I Program," in *Robert Moses and the Modern City: The Transformation of New York*, ed. Hilary Ballon and Kenneth T. Jackson (New York: W. W. Norton), 65.

3. Thomas J. Shelley, *The Bicentennial History of the Archdiocese of New York: 1808–2008* (Strasbourg: Éditions du Signe, 2008), 512. Likewise, Governor Nelson Rockefeller addressed Spellman as "Your Eminence," while to Spellman the governor was "Nelson." See letter from Spellman to Rockefeller, December 7, 1963, NAR RG4 Personal Series L, box 235, folder 2361. I am indebted to Monica Blank, archivist at the Rockefeller Archives Center, for her assistance.

4. Spellman's relationship with trade unions was not always congenial. In 1949, when gravediggers at two Catholic cemeteries went on strike, Spellman enlisted seminarians to break it. Thomas J. Shelley, "Francis J. Spellman: The Evolution of an Ecclesiastical Career," *American Catholic Studies* 114, no. 3 (Fall 2003): 90, and Charles J. V. Murphy, "The Cardinal and the City," *Fortune*, February 1960, 183–84.

5. "Roman Catholics: The Pastor-Executive," *Time*, May 15, 1964.

6. Robert Caro, *The Power Broker: Robert Moses and the Fall of New York* (New York: Knopf, 1974), 741.

7. Lawrence McCaffrey, *The Irish Diaspora in America* (Bloomington: Indiana University Press, 1976), 151. For this quote I am indebted to Thomas J. Shelley, *Bicentennial History of the Archdiocese of New York*, 27.

8. "Civitas Dei," *Tablet* (London) January 28, 1958, 80. I am grateful to Bill Cotter for the fuller translation of the theme of the fair.

9. "Civitas Dei," 80.

10. "Setting the Stage for the Brussels Fair," *New York Times*, March 16, 1958, 2.

11. Howard Fussiner, "Art at the Fair," *College Art Journal* 18, no. 1 (Autumn 1958): 68–71.

12. Howard Taubman, "The Arts: A Critic's View," *New York Times*, April 22, 1964, 24.

13. Fussiner, "Art at the Fair," 68–71.

14. "Americans at Brussels: Soft Sell, Range & Controversy," *Time*, June 16, 1958, 71. Marguerite Cullman, the wife of the commissioner of the US pavilion, Howard Cullman, compiled her behind-the-scenes impressions of the fair in her book *Ninety Dozen Glasses* (W. W. Norton, 1960).

15. Fussiner, "Art at the Fair," 68–71.

16. "Americans at Brussels," *Time*, June 16, 1958.

17. Report on the Vatican pavilion, "Universal International Exhibition at Brussels: April 15, 1958–October 15, 1958," by Bishop Russell J. McVinney of Providence, RI, box S-C4, folder 9, Archdiocese of New York Archives (hereafter AANY).

18. "Civitas Dei," 8.

19. "Civitas Dei," 8.

20. These popes were Pius XII, John XXIII, Paul VI, and John Paul II. From the Arthur Fleischmann official website: http://www.fleischmann.org.uk/frbiog.html.

21. Fleischmann pioneered the use of Perspex, a form of plexiglass, for sculpture.

22. "Civitas Dei," 8.

23. Massimo Faggioli, *John XXIII: The Medicine of Mercy* (Collegeville, MN: Liturgical, 2014), 112.

24. "The Bob Moses World's Fair," *Herald Tribune*, March 3, 1960, 16.

25. John Canaday, "Happy New Year! Thoughts on Critics and Certain Painters as the Season Opens," *New York Times*, September 6, 1959.

1. The Vision

1. Robert Moses, "From Dump to Glory," *Saturday Evening Post*, January 15, 1938, 13.

2. Robert Moses, *The Saga of Flushing Meadow, the Valley of Ashes* (New York: Triborough Bridge and Tunnel Authority, 1966), 1.

3. Caro, *Power Broker*, 10.

4. Galen Cranz, *The Politics of Park Design: A History of Urban Parks in America* (Cambridge, MA: MIT Press, 1982), 61.

5. Cranz, 63.

6. Cranz, 102.

7. Caro, *Power Broker*, 243.

8. Cranz, *Politics of Park Design*, 106.

9. Cranz, 108–9.

10. Roberta Brandes Gratz, *The Battle for Gotham: New York in the Shadow of Robert Moses and Jane Jacobs* (New York: Nation Books, 2010), 132.

11. Paul Goldberger, "The Urban Planner in Spite of Himself," *New York Times*, August 2, 1981.

12. Caro, *Power Broker*, 12.

13. Paul Goldberger, "Robert Moses, Master Builder, Is Dead at 92," *New York Times*, July 30, 1981.

14. Gratz, *Battle for Gotham*, 126–27.

15. Paul Goldberger, "Eminent Dominion: Rethinking the Legacy of Robert Moses," *New Yorker*, February 5, 2007, 85, https://www.newyorker.com/magazine/2007/02/05/eminent-dominion.

16. Moses, "From Dump to Glory," 72.

17. Caro, *Power Broker*, 1084.

18. Caro, 1085.

19. Marc H. Miller, "Something for Everyone: Robert Moses and the Fair," in *Remembering the Future: The New York World's Fair from 1939–1964*, ed. Robert Rosenblum (Queens Museum, New York, and Rizzoli International, 1989), 56.

20. Jesse T. Todd, "Imagining the Future of American Religion at the New York World's Fair, 1939–40" (PhD diss., Columbia University, 1996), 6.

21. Todd, 15.

22. "Temple of Religion Called Key to the Fair," *New York Times*, January 10, 1939.

23. Todd, "Imagining the Future," 215.

24. Miller, "Something for Everyone," 56.

25. Caro, *Power Broker*, 333 n.

26. Samuel, *End of the Innocence*, 6.

27. Dan Magnan, "Shea Was His Field of Dreams—Moses Envisioned It as Jewel of Complex," *New York Post*, October 24, 2000.

28. Joseph Tirella, *Tomorrow-Land: The 1964–65 World's Fair and the Transformation of America* (Guilford, CT: Lyons, 2013), 3.

29. Martin J. Manning, "Fairs! Fairs! Fairs! The United States Information Agency and U.S. Participation at World Fairs since World War II," *Popular Culture in Libraries* 2, no. 3 (1994): 4.

30. Manning, 4.

31. Seattle's 1962 World's Fair was classified as an "international" fair.

32. John Brooks, "Diplomacy at Flushing Meadow," Onward and Upward with the Arts, *New Yorker*, June 1, 1963, 42. Also referenced, Jack Masey and Conway Lloyd Morgan's *Cold War Confrontations: US Exhibitions in the Cultural Cold War* (New York: Lars Müller, 2008), 110.

33. Manning, "Fairs! Fairs! Fairs!," 4–5.

34. Manning, 7, 29, n. 4.

35. I am grateful to Bill Cotter for emphasizing the potential cost to countries to participate in the New York fair.

36. Martin Mayer, "Ho Hum, Come to the Fair," *Esquire*, October 1963, 379.

37. "Moses Dismisses Criticism of the Fair," *New York Times*, September 11, 1963.

38. Charles Poletti, "The Reminiscences of Charles Poletti," vol. 4, 1978, 568–69, Oral History Research Office, Columbia University, New York. In recalling the 1960 papal meeting, Poletti said that "whenever I went into the Vatican all the people knew me because I'd been there several times during the war, so they all came up to say hello to me and so forth, and I had with me a very distinguished Catholic layman, and I think he was a little surprised that little Protestant Poletti was rated so high in the Vatican."

39. Brooks, "Diplomacy at Flushing Meadow," 42.

40. Samuel, *End of the Innocence*, xx.

2. Francis Cardinal Spellman

1. Charles R. Morris, *American Catholic: The Saints and Sinners Who Built America's Most Powerful Church* (New York: Random House, 1997), 219.

2. Alden Whitman, "Francis J. Spellman: New York Archbishop and Dean of American Cardinals," *New York Times*, December 3, 1967, 82.

3. Robert I. Gannon, *The Cardinal Spellman Story* (New York: Doubleday, 1963), 47.

4. The Diocese of Brooklyn, over which Cardinal Spellman had no practical authority, includes Queens and Brooklyn.

5. Murphy, "Cardinal and the City," 152–54.

6. "Cardinal Spellman: Symbol of His Age," *Reconstructionist*, December 29, 1967, 5.

7. "The Year in Books," *Time*, December 18, 1950. Spellman's relationship with Hollywood was adversarial. Spellman lost media support for his criticism of Hollywood's relaxed moral standards. The Catholic Church did not have the power to censor films that had objectionable content, but it did have the power to influence. In line with the Motion Picture Production Code, the Catholic-supported Legion of Decency was established to review and rate film content according to Catholic teachings. Spellman also used his close friendship with Louis B. Mayer ("L.B." to Spellman), cofounder of Metro-Goldwyn-Mayer, to press a point. Likewise, Mayer would call on Spellman to soften a film's rating. See Neal Gabler, *An Empire of Their Own: How Jews Invented Hollywood* (New York: Anchor Books, 1988), 286. Ironically, feminist academic and social critic Camille Paglia credits these censorial efforts for "Hollywood's supreme era." See Paglia, "Camille Paglia on Movies, #MeToo and Modern Sexuality: 'Endless, Bitter Rancor Lies Ahead,'" *Hollywood Reporter*, February 27, 2018, https://www.hollywoodreporter.com/news/camille-paglia-movies-metoo-modern-sexuality-endless-bitter-rancor-lies-1088450.

8. Peter Hebblethwaite, *Pope John XXIII: Shepherd of the Modern World; The Definitive Biography of Angelo Roncalli* (Garden City, NY: Image Books, 1987), 462.

9. George A. Hoehmann, "'My Eminent Friend of New York': Francis Cardinal Spellman and the Second Vatican Council" (Master's thesis, St. Joseph's Seminary, 1992).

10. Hoehmann, 57.

11. Avery Dulles, *A Testimonial to Grace and Reflections on a Theological Journey* (Kansas City, MO: Sheed & Ward, 1996), 109.

12. Thomas J. Shelley, *The Bicentennial History of the Archdiocese of New York* (Strasbourg: Éditions du Signe, 2008), 282.

13. Hoehmann, "My Eminent Friend," 154.

14. Edward Penton, "Why Did Vatican II Ignore Communism?," *Catholic World Report*, December 12, 2010, https://www.catholicworldreport.com/2012/12/10/why-did-vatican-ii-ignore-communism.

15. Francis Spellman, "Communism Is Un-American," *American Magazine*, July 1946, 26–28.

16. Spellman, 26.

17. Donald F. Crosby, *God, Church, and Flag: Senator Joseph R. McCarthy and the Catholic Church, 1950–1957* (Chapel Hill: University of North Carolina Press, 1978), 14.

18. Gannon, *Cardinal Spellman Story*, 444.

19. John E. Haynes, *Red Scare or Red Menace?* (Chicago: Ivan R. Dee, 1996), 93.

20. Thomas J. Shelley, "Slouching toward the Center: Cardinal Francis Spellman, Archbishop Paul J. Hallinan and American Catholicism in the 1960s," *U.S. Catholic Historian* 17, no. 4 (Fall 1999): 23–49.

21. B. Katolin, "The Servant of Three Gentlemen," *Trud*, no. 29 (May 31, 1961), translation March 29, 1981, 5.

22. Shelley, *Bicentennial History*, 530.
23. Shelley, 531.
24. Shelley, 520.
25. Murphy, "Cardinal and the City," 61. The cardinal's construction projects were valued, at the time of his death, at more than a half-billion dollars and spread over an archdiocese of 4,717 square miles.
26. Caro, *Power Broker*, 741.
27. "Urban Renewal Unit Formed in New York," *Catholic Standard and Times*, May 19, 1961, 13.
28. "Fordham's History: A Transformative History," Fordham University, https://www.fordham.edu/about/fordhams-history/.
29. Hilary Ballon and Kenneth T. Jackson, eds., *Robert Moses and the Modern City: The Transformation of New York* (New York: W. W. Norton), 201.
30. Spellman to Moses, December 15, 1955, Moses to Spellman, December 22, 1965, NYPL-CSC, box 116.
31. Hilary Ballon, "Robert Moses and Urban Renewal: The Title 1 Program," in Ballon and Jackson, *Robert Moses and the Modern City*, 94–96, 107.
32. Shelley, *Bicentennial History of the Archdiocese*, 282.

3. Moses and Spellman

1. Arnaldo Cortesi, "Vatican Hints Role in 1964 Fair Here," *New York Times*, September 4, 1960.
2. Poletti, "Reminiscences," 4:567.
3. Tardini died in Rome on July 30, 1961, of a massive heart attack and was replaced by Amleto Giovanni Cicognani.
4. Cardinal Spellman, letter to bishops, undated, 1964–1965 World's Fair Collection, Collection Number 025.002, box 95, folder 5, AANY.
5. Lotus Club, https://www.lotosclub.org.
6. W. E. Potter to Robert Moses, memorandum, December 8, 1960, box 134, New York World's Fair 1964–1965 Corporation records, Archives and Manuscripts Division, New York Public Library (hereafter NYPL-NYWF).
7. Howard Taubman, "The Arts: A Critic's View," *New York Times*, April 22, 1964, 24.
8. Robert Moses to Thomas Deegan, memorandum, April 2, 1961, box 134, NYPL-NYWF.
9. Samuel, *End of the Innocence*, 21.
10. Thomas Deegan to Robert Moses, memorandum, August 8, 1961, box 134, NYPL-NYWF.
11. Deegan to Moses, August 8, 1961.
12. Thomas Deegan to Charles Poletti, November 8, 1961, box 134, NYPL-NYWF.
13. John Young to Robert Moses, memorandum, October 10, 1962, box 134, NYPL-NYWF.
14. John J. Maguire to Walter J. Lesch, January 11, 1965, box SC92, folder 1, AANY.
15. Robert Moses to Thomas J. Deegan, April 28, 1961, box 134, NYPL-NYWF.
16. Moses to Deegan, April 28, 1961.
17. Roland Redmond to Robert Moses, March 19, 1962, box 134, NYPL-NYWF.
18. Redmond to Moses, March 19, 1962. This is exactly how the designer Jo Mielziner staged the sculpture. There is no evidence that this was communicated to Mielziner; however, it could have easily come up in discussions. On the other hand, as an experienced artist, he may have independently envisioned the same approach.

19. Edward M. Kinney, *The Saga of a Statue* (self-published, 1989), 6.

20. Algase was also the literary agent for John F. Kennedy's *When England Slept* and is credited with originating the Alfred F. Smith Foundation Dinner in 1945 to raise money for the city's poor. Gertrude Algase, a Literary Agent, *New York Times*, August 4, 1962, 19.

21. Gay Talese, "Fair Sees 'Pieta' as Top Feature," *New York Times*, April 11, 1962, 45.

22. Josef Vincent Lombardo, *Michelangelo: The Pietà and Other Masterpieces* (New York: Pocket Books, 1965), 50.

23. "Florentines Assail Loan of Art to U.S.," *New York Times*, October 15, 1956, 27.

24. "4 Artists in Barricade Fight Art Loan to U.S.," *New York Times*, October 25, 1956, 36.

25. Charles J. Poletti to Robert Moses, memorandum, April 9, 1962, box 134, NYPL-NYWF.

26. "Rome Letter," *Tablet*, April 7, 1962, 7.

27. "Rome Letter."

28. "Rome Letter."

29. Kinney, *Saga of a Statue*, 9.

30. Talese, " 'Fair Sees 'Pieta' as Top Feature," 45.

31. "Shipping of Pieta Queried," Letters to the Times, *New York Times*, April 23, 1962, 27.

32. Alfred Frankfurter, "The Michelangelo Scandal," *ARTnews*, May 1962, 6.

33. Frankfurter, 6.

34. John McCloy to Cardinal Spellman, April 12, 1962, box 90, folder 4, AANY.

35. "Exhibiting the 'Pieta,' " Letters to the Times, *New York Times*, April 30, 1962, 25.

36. "Bringing 'Pieta' Opposed," Letters to the Times, *New York Times*, May 7, 1962, 30.

37. "Exhibiting 'Pieta' Opposed," Letters to the Times, *New York Times*, May 25, 1962, 32.

38. Robert Moses to John Coolidge, May 21, 1962, box 134, NYPL-NYWF.

39. "Vatican to Send Statues to Fair," *New York Times*, April 17, 1963, 43.

40. Herman Lebovics, *Mona Lisa's Escort: André Malraux and the Reinvention of French Culture* (Ithaca, NY: Cornell University Press, 1999), 9.

41. Margaret Leslie Davis, *Mona Lisa in Camelot* (New York: Da Capo, 2008), 71.

42. Davis, 95.

43. Davis, 59.

44. Emanuel Perlmutter, "World's Fair Bid to Peiping Barred," *New York Times*, June 3, 1962, 41.

45. Peter Hebblethwaite, *Paul VI: The First Modern Pope* (New York: Paulist, 1993), 327.

46. "Pope Is Reported Hesitating on Plan for Pieta at Fair," *New York Times*, September 3, 1962; "Reports of Vatican Shift on La Pieta Called 'Gossip,' " *New York Times*, September 4, 1962, 6.

47. "Greece May Send Hermes to '64 Fair," *New York Times*, September 13, 1963, 24.

48. "Hermes at Fair Approved," Letters to the Times, *New York Times*, December 2, 1963, 36.

49. "Venus de Milo Returned to France Safe and Sound," *New York Times*, August 2, 1964, 53.

50. "Venus, a Bit Damaged, Visits Japan," *New York Times*, March 23, 1964, 22.

51. Alfred Frankfurter, "On Making the World Safe for Art," *ARTnews*, February 1964, 23.

52. Marianna Hassol, "Is This Trip Necessary?," *Boston Globe*, March 22, 1964.

53. Frankfurter, "On Making the World Safe for Art," 23.

54. Robert Allen, " 'Pieta' Unveiled in Glow of Blue," *New York Times*, April 20, 1964, 1.

55. C. J. McNaspy, "World's Fair Preview," *America*, April 11, 1964, 510.

56. Hassol, "Is This Trip Necessary?," 212.

4. The Coup

1. Lombardo, *Michelangelo*, 27.
2. Lombardo, 13.
3. Lombardo, 14.
4. Charles de Tolnay, *Youth of Michelangelo* (Princeton, NJ: Princeton University Press, 1947), 146.
5. Lombardo, *Michelangelo*, 48.
6. Lombardo, 29.
7. Kathleen Wiel-Garris Brandt, "Michelangelo's Pietà for the Cappella del Re di Francia," in *Michelangelo: Selected Scholarship in English (Book 1)*, ed. William E. Wallace (New York: Routledge, 1996), 220.
8. Barbara G. Lane, *The Altar and the Altarpiece: Sacramental Themes in Early Netherlandish Painting* (New York: Harper & Row, 1984), 35.
9. Tolnay, *Youth of Michelangelo*, 147.
10. Frederick Hartt, *History of Italian Renaissance Art: Painting, Sculpture, Architecture* (New York: Harry N. Abrams, 1979), 477.
11. Hartt, 477.
12. Frederick Hartt, *Michelangelo: The Complete Sculpture* (New York: Harry N. Abrams, 1968), 83.
13. Hartt, *History of Italian Renaissance Art*, 477.
14. Hartt, 477.
15. Tolnay, *Youth of Michelangelo*, 150.
16. Lombardo, *Michelangelo*, 114.
17. Giorgio Vasari, *The Lives of the Artists*, trans. Julia Conway Bondanella and Peter Bondanella, Oxford World Classics (New York: Oxford University Press), 1991, 425.
18. Gillian Vallance Mackie, *Early Christian Chapels in the West: Decoration, Function and Patron* (Toronto: University of Toronto Press, 2003), 59.
19. Louise Rice, *The Altars and Altarpieces of New St. Peter's: Outfitting the Basilica, 1621–1666* (Cambridge: Cambridge University Press, 1997), 52–53.
20. Vasari, *Lives of the Artists*, 425; Brandt, "Michelangelo's Pietà," 232.
21. Rice, *Altars and Altarpieces*, 216.
22. Rice, 216.
23. Rice, 217.
24. Grżyna Jurkowlaniec, "A Miracle of Art and Therefore a Miraculous Image: A Neglected Aspect of the Reception of Michelangelo's Vatican Pietà," *Artibus et Historiae* 36, no. 7 (2015): 189. Considered a distraction, the putti with the crown were removed in 1927.
25. Thomas Craven, *Men of Art* (New York: Simon & Schuster, 1940), 126.
26. "Fair Will Revise Display of Pieta," *New York Times*, February 23, 1964, 59.
27. Grove Art, online reference, https://www-oxfordartonline-com.
28. For two centuries, the Palestrina *Pietà* was also attributed to Michelangelo, but scholars have cast doubt on its authenticity, complicated in part by the fact that this sculpture also is unfinished. It is a High Renaissance sculpture, but the sculptor has not been identified.

5. Spellman's Dream Team

1. Albin Krebs, "Howard S. Cullman, 80, of Port Authority, Dies," *New York Times*, June 30, 1972.
2. Howard S. Cullman to Msgr. Charles J. McManus, St. Patrick's Cathedral, November 14, 1960, box SC90, folder 1, AANY.

3. Cullman to McManus, November 14, 1960.

4. Charles Poletti to John Young, memorandum, July 25, 1962, box 134, NYPL-NYWF.

5. Equivalent to $32 million in 2020.

6. Timothy J. Flynn, "The Vatican Pavilion," *Catholic Market*, January 1964, 10.

7. "Archbishop Bryan McEntegart, Bishop of Brooklyn, 75, Is Dead; Prelate Headed the Largest Catholic Diocese in U.S. from 1957 to 1968," *New York Times*, October 1, 1968, 48.

8. Edward Tivana, "A New Yorker Up for Sainthood," *New York Times*, November 11, 1986.

9. "Set Dedication of Vatican Pavilion at World's Fair," *Tablet*, October 10, 1963, Msgr. James Asip Collection, Vatican Pavilion, box 1, Roman Catholic Diocese of Brooklyn, Diocesan Archives, Office of the Bishop (hereafter RCDBA).

10. "Pope's Signal Starts Pavilion at Fair," *New York Times*, November 1, 1962, 33.

11. Msgr. Timothy Flynn to Vatican Pavilion Directors, memorandum, January 18, 1963, Project Files, 1933–1966, Vatican Pavilion, box 1, RCDBA.

12. Flynn, "The Vatican Pavilion," *Catholic Market*, January 1964, 11, RCDBA.

13. "IBM Pavilion NY World's Fair," Eames Office, https://www.eamesoffice.com/the-work/ibm-pavilion-ny-worlds-fair/.

14. Bishop McEntegart to Cardinal Spellman, March 18, 1963, Vatican Pavilion, box 2, RCDBA.

15. Director meeting minutes, March 22, 1963, Vatican Pavilion, box 2, RCDBA. It was also reported that for the musicians alone, it would require approximately thirty-five musicians at a minimum cost of $26 per musician, plus $5 per hour for rehearsal time, to produce one High Mass at the World's Fair Chapel, according to Local 802.

16. Director meeting minutes, March 14, 1963, box SC91, folder 4, AANY.

17. Vatican Pavilion, box 3, RCDBA.

18. Meeting minutes, April 8, 1963, box SC91, folder 4, AANY.

19. Vatican Pavilion Exhibit Report on meetings and progress of exhibit, box SC91, folder 4, AANY.

20. Director meeting minutes, Vatican Pavilion, box 1, RCDBA.

21. Msgr. Asip to Bishop McEntegart, meeting minutes, May 29, 1963, Vatican Pavilion, box 1, RCDBA.

22. Review of contributors, box 94, folder 2, AANY.

23. "Catholics Asked to Support Vatican Pavilion at Fair," *New York Times*, May 20, 1963, 49.

24. Review of contributors, box 94, folder 2, AANY.

25. Msgr. Asip to Bishop McEntegart, meeting minutes, May 29, 1963, Vatican Pavilion, box 1, RCDBA.

26. John Young to Robert Moses, memorandum, December 6, 1963, box 134, NYPL-NYWF.

27. Morris J. MacGregor, *Steadfast in the Faith* (Washington, DC: Catholic University of America Press, 2005), 89; Joseph McLellan, "Catholic Relief Services Has World's Largest First Aid Kit," *Pittsburgh Catholic*, March 3, 1972, 8; National Catholic News Service, Wednesday, September 1, 1976.

28. Kinney, *Saga of a Statue*, 7.

29. Kinney, 7.

30. George Weller, "Sub Due to Bring 'Pieta' of Michelangelo to U.S.," *Washington Post*, October 20, 1962, C4.

31. My gratitude to Bill Cotter for this insight.

32. Elizabeth J. Milleker, "A Brief History of the Cast Collection at the Metropolitan Museum of Art," monograph, Metropolitan Museum of Art, n.d.

33. Milleker, "Brief History."

34. Douglas C. McGill, "Plaster Casts of Statues: From Storage into Vogue," *New York Times*, January 1, 1987.

35. Kinney, *Saga of a Statue*, 8.

36. Kinney, 15.

37. Kinney, 30.

38. James B. Gordon, "Packing of Michelangelo's 'Pieta,'" *Studies in Conservation* 12, no. 2 (May 1967): 57.

39. Gordon, 58.

40. Gordon, 58.

41. Gordon, 59.

42. Kinney, *Saga of a Statue*, 30–31.

6. Moving Marble

1. Kinney, *Saga of a Statue*, 8.

2. William E. Wallace, "An Impossible Task," in *Making and Moving Sculpture in Early Modern Italy*, ed. Kelley Helmstutler Di Dio (Burlington, VT: Ashgate, 2015), 48.

3. Robert G. Rosegrant, "Packing Problems and Procedures," *Technical Studies in the Field of the Fine Arts* 10, no. 3 (1942): 142.

4. Robert P. Sugden, *Care and Handling of Art Objects* (New York: Metropolitan Museum of Art, 1946), 11–12.

5. Sugden, 11.

6. Kinney, *Saga of a Statue*, 15.

7. Kinney, 17.

8. Galeazzi designed the present North American College in Rome. The count is entombed in the crypt in the college's chapel. http://orbiscatholicussecundus.blogspot.com/2010/11/layman-at-service-of-holy-see-enrico.html.

9. Thomas Maier, *When Lions Roar: The Churchills and the Kennedys* (New York: Crown, 2014), 368.

10. Kinney, *Saga of a Statue*, 18.

11. "Risk Management & Audit Services," Harvard University, https://rmas.fad.harvard.edu/pages/goods-equipment-transit-risk-discussion.

12. Kinney, *Saga of a Statue*, 18.

13. Kinney, 21.

14. Vatican Office to Charles Poletti, May 21, 1962, box 134, folder 1, NYPL-NYWF.

15. Kinney, *Saga of a Statue*, 21.

16. John P. Callahan, "Pieta to Get Massive Protection: Statue Insured for $10 Million for Its Trip to Fair," *New York Times*, January 16, 1964, 33.

17. Kinney, *Saga of a Statue*, 23.

18. Kinney, 25.

19. Kinney, 36.

20. Kinney, 25.

21. Kinney, 26.

22. "X-rays Show Repairs to Broken Pieta," *Catholic Courier Journal*, April 23, 1964, 1.

23. Kinney, *Saga of a Statue*, 40.

24. Kinney, 41.

25. "Dispute over Packing May Hold Up Pieta," *Boston Globe*, March 26, 1964, 10.

26. John S. Young to [Charles] Poletti, memorandum, April 13, 1964, box 286, NYPL-NYWF.

27. "Venus Prepared for Trip to Japan," *New York Times*, February 7, 1964, 28.
28. John Young to Charles Poletti, telegram, March 24, 1964, box 286, NYPL-NYWF.
29. Kinney, *Saga of a Statue*, 44.
30. Young to Poletti, 3.
31. Young to Poletti, 4.
32. Kinney, *Saga of a Statue*, 45.
33. "Packing the Pieta Revisited: American Know-How Paid Off," Catholic News Service—Newsfeeds, June 2, 1972.
34. Kinney, *Saga of a Statue*, 48.
35. Meeting minutes, April 8, 1963, box SC91, folder 4, AANY.
36. "Pieta Moved off Pedestal," *New York Times*, April 2, 1964, 30.
37. Kinney, *Saga of a Statue*, 51.
38. Kinney, 51.
39. Kinney, 64.
40. Kinney, 56.
41. Kinney, 58.
42. My gratitude to Bill Cotter, whose father had firsthand knowledge of these security precautions as a peer review consultant for the US shipping team.
43. Kinney, *Saga of a Statue*, 60.
44. Kinney, 63.
45. Robert Alden, "The 'Pieta' Arrives Here, Ever So Gently: Michelangelo Statue Lifted from Liner for Trip to Fair," *New York Times*, April 14, 1964, 27.
46. Kinney, *Saga of a Statue*, 64.
47. Kinney, 65.
48. Kinney, 66.
49. Vatican Pavilion, *New York World's Fair 1964–1965: A Chronicle* (New York: New York World's Fair), 1966, 11.

7. On Stage

1. Mary C. Henderson, *Mielziner: Master of Modern Stage Design* (New York: Backstage Books, 2001), 6.
2. Albin Krebs, "Jo Mielziner Dead at 74; Pioneering Set Designer," *New York Times*, March 16, 1976, 38.
3. Jo Mielziner, *Designing for the Theatre: A Memoir and a Portfolio* (New York: Atheneum, 1965), 3, 243.
4. Milton Lomask, "Return to Orthodoxy," *Sign*, January 1955, 31.
5. Kristine Somerville, "Stage Pictures: Jo Mielziner and the Art of Set Design," *Missouri Review* 42, no. 3 (Fall 2019): 62.
6. Henderson, *Mielziner*, 302.
7. "He Sets the Stage," *Catholic Digest* 65, no. 2 (July 1962).
8. Henderson, *Mielziner*, 105.
9. Henderson, 85.
10. Henderson, 243.
11. Henderson, 96.
12. Henderson, 96.
13. Lomask, "Return to Orthodoxy," 31.
14. "He Sets the Stage."
15. Henderson, *Mielziner*, 97–98.

16. Krebs, "Jo Mielziner Dead at 74."

17. Henderson, *Mielziner*, 135.

18. Stanley J. Kusman, obituary, press release issued April 17, 1990, from St. Mary's University, San Antonio, Texas.

19. Henderson, *Mielziner*, 135. Soldiers on the front lines on night patrol were being killed and left without a decent burial. Under no obligation to do so, Chaplain Kusman volunteered to retrieve their bodies and went out alone into the no-man's-land between the lines and carried on his shoulders the bodies of fifty-seven soldiers—forty-seven Americans and ten Germans—and buried them all. For this act of valor, Kusman received a Purple Heart and was nominated for a Bronze Star. Mielziner's friendship with Kusman continued after the war ended, and in 1974 Mielziner traveled to San Antonio to speak at Kusman's golden jubilee in the church.

20. Lomask, "Return to Orthodoxy," 31.

21. Henry Artell, "'I Looked at It with Awe'—as Thousands Do Daily: An Interview with Jo Mielziner," *Catholic Courier Journal*, May 7, 1964, 1.

22. Jo Mielziner to Archbishop Spellman, April 12, 1962, box SC90, folder 4, AANY.

23. Richard F. Shepard, "Show Biz and Big Biz Are Offering Entertainment," *New York Times*, April 22, 1964, 24. Mielziner also produced the Bell Telephone exhibit, which was a multimedia history of communications.

24. Mielziner to Charles de Tolnay, January 4, 1963, Jo Mielziner Papers, box 15, folder 8, *T-Mss 1993–002, Billy Rose Theatre Division, New York Public Library for the Performing Arts (hereafter BRTD-NYPLPA).

25. Vatican Pavilion press release, Jo Mielziner's description of his setting for the *Pietà* at the Vatican Pavilion, undated, Msgr. James Asip Collection, Vatican Pavilion, box 1, RCDBA.

26. List and description of pavilion exhibits, Msgr. James Asip Collection, Vatican Pavilion, box 1, RCDBA.

27. Frederick Voss to Mielziner, July 16, 1962, box 15, folder 8, BRTD-NYPLPA.

28. Artwell, "'I Looked at It with Awe,'" 1.

29. Rev. Heneghan to Msgr. Cooke, October 29, 1962, 1964–1965 World's Fair Collection, AANY.

30. Mielziner to William C. Jaeger Jr., February 21, 1963, box 15, folder 8, BRTD-NYPLPA.

31. Mielziner to Glbe Derujinsky, contract for $150, December 20, 1962, box 15, folder 8, BRTD-NYPLPA.

32. Pavilion directors' meeting minutes, March 7, 1963, Msgr. James Asip Collection, Vatican Pavilion, box 1, RCDBA.

33. Mielziner to Rev. John J. Maguire et al., memorandum, March 14, 1963, re "Protective Glass for 'Pieta,'" box SC 90, folder 7, AANY.

34. Mielziner, "Notes on Investigation re Transparent Screen for 'Pieta,'" March 30, 1963, Msgr. James Asip Collection, Vatican Pavilion, box 1, RCDBA.

35. Vatican Pavilion Progress Report #3, July 1, 1963, box 93-F3, AANY.

36. Giuseppe Tommasi Studios budget proposal, October 15, 1963, box 70, BRTD-NYPLPA.

37. Peter Feller to Jo Mielziner, pavilion estimate, August 13, 1963, box 70, BRTD-NYPLPA.

38. "'Staging' the Pieta," *Denver Catholic Register*, August 25, 1963, box 16, folder 5, BRTD-NYPLPA.

39. Irving Stone to Jo Mielziner, May 23, 1963, box 16, folder 5, BRTD-NYPLPA.

8. Moses Delivers

1. "Ground Broken at Fair—President Wears Topcoat," *New York Times*, December 15, 1962, 8.

2. "'Stall-In' Scored by Negro Leaders," *New York Times*, April 17, 1964, 18.

3. "Johnson Critical of Fair Chanting," *New York Times*, April 24, 1964, 1.

4. Robert Moses, "From Dump to Glory," *Saturday Evening Post*, January 15, 1938, 13.

5. Adam Rogers, "Global Ambition," Smithsonian, June 2017, 12.

6. Samuel C. Florman, "Deus ex Machina," *Mechanical Engineering* 134, no. 12 (December 2012): 42–45. After the building was razed in 1976, the site would become the National Tennis Center's Arthur Ashe Stadium twenty-five years later. Samuel, *End of the Innocence*, 195.

7. Samuel, *End of the Innocence*, 126–27.

8. Samuel, 110.

9. Robert Moses, *Public Works: A Dangerous Trade* (New York: McGraw-Hill, 1970), 586.

10. "Legacy of Robert Moses Hailed," *New York Times*, August 1, 1981, 29.

11. Paul Montgomery, "Religion Present throughout the Fair: 8 Pavilions Add Depth to Spirit of Carnival," *New York Times*, April 22, 1964, 25.

12. Elizabeth Macaulay-Lewis and Jared Simard, "From Jerash to New York: Columns, Archaeology, and Politics at the 1964–65 World's Fair," *Journal of the Society of Architectural Historians* 74, no. 3 (September 2015): 355; *1965 Official Guide: New York World's Fair* (New York: Time-Life Books, 1965), 122, 147.

13. Macaulay-Lewis and Simard, "From Jerash to New York," 360. A gift from King Hussein to the fair, the Column of Jerash, remains on the site. It, too, was not without controversy, as it was not the column originally promised—it was shorter. Concerned that Moses might not accept it, fair heads Charles Poletti and Lionel Harris did not reveal the substitution.

14. "United Protestant Center Planned at World's Fair," *New York Times*, October 27, 1962.

15. Julie Nicoletta, "Selling Spirituality and Spectacle: Religious Pavilions at the New York World's Fair of 1964–1965," *Buildings & Landscapes: Journal of the Vernacular Architecture Forum* 22, no. 2 (Fall 2015): 62–68.

16. Montgomery, "Religion Present throughout the Fair," 25.

17. Bernard Weinraub, "Thousands Combine Worship with a Visit to the World's Fair," *New York Times*, April 27, 1964, 20.

18. Weinraub, 20.

19. New York State Library, "Welcome to the Fair! The 1939 and 1964 New York World's Fairs," https://www.nysl.nysed.gov/collections/worldsfair/.

20. *New York Times*, "'Parable' Draws Crowds at Fair; Disputed Film Helps to Pay for Protestant Center," August 7, 1964, 14.

21. The Protestant Council of the City of New York had received rental or purchase orders predominately from Protestant sources, with only sporadic interest from Catholics. Then all of a sudden a large number of orders started coming in from Catholics, such as Holy Name societies, CYO, and CCD groups. This was attributed to a five-page, single-spaced typed commentary titled "A Suggested Way of Using 'Parable' in a Closed Retreat," written by a Jesuit, Father Thomas Gedeon, of the Jesuit Retreat House in Cleveland. The film's distributor, George F. Geiss, began including the commentary with all orders placed by Catholic buyers. He said "I don't send it to other religious groups that way, because they'd wonder why we were pushing the Catholic commentary, but if anyone asks me for a discussion guide, this

is what I suggest to them, and they're usually grateful for it." *National Catholic Reporter*, May 11, 1966, 9.

22. R. Scott Lloyd, "1964 World's Fair Had Far-Reaching Impact," *Deseret News*, September 30, 2014, https://www.deseret.com/2014/9/30/20549559/1964-world-s-fair-pavilion-had-far-reaching-impact.

23. Lloyd, "1964 World's Fair."

24. *1965 Official Guide*, 156.

25. *1965 Official Guide*, 106.

26. *1965 Official Guide*, 132, 137.

27. "Nubian Treasure Arriving for Fair," *New York Times*, April 16, 1964, 22; "Sudan's Madonna Damaged at Fair," *New York Times*, June 26, 1964, 18.

28. Arthur P. Molella, "The Human Spirit in an Age of Machine: The Pietà and the Computer at the 1964–1965 New York World's Fair," in *World's Fairs in the Cold War: Science, Technology, and the Culture of Progress*, ed. Arthur P. Molella et al. (Pittsburgh: University of Pittsburgh Press, 2019), 101.

29. Molella, 101.

30. Michele H. Bogart, *Sculpture in Gotham: Art and Urban Renewal in New York City* (London: Reaktion Books, 2018), 33–34.

31. John Canaday, "Art All Over the Fair," *New York Times*, 1964.

32. Suzanne P. Fredericks, *Marshall M. Fredericks, Sculptor* (Saginaw, MI: Saginaw Valley State University, 2003), 152.

33. Robert Beverly Hale, "The American Moderns," *Metropolitan Museum of Art Bulletin*, n.s., 16, no. 1 (Summer 1957): 20.

34. Douglas C. McGill, "José De Rivera Is Dead at 80; Known for Metal Sculptures," *New York Times*, March 21, 1985.

35. Wayne Craven, *Sculpture in America* (London: Cornwall Books, 1984), 632.

36. Bernard Holland, "Theodore Roszak, Sculptor of Eagle at British Embassy" (obituary), *New York Times*, September 4, 1981, 12.

37. Bogart, *Sculpture in Gotham*, 39.

38. After the fair closed, Moses commissioned De Lue to cast in bronze a replica of his plaster cast of *George Washington as Master Mason*, completed for the Masonic pavilion. The bronze is now a permanent monument at the park.

39. John Canaday, "Pardon the Heresy," *New York Times*, March 29, 1964, 19.

40. Joseph Lelyveld, "Moses Is Accused of Barring Art," *New York Times*, March 2, 1964, 1.

41. Philip Benjamin, "Fine Arts Center Opens at Fair," *New York Times*, June 23, 1964, 19.

42. Samuel, *End of the Innocence*, 66.

43. Joseph Tirella, *Tomorrow-Land: The 1964–65 World's Fair and the Transformation of America* (Guilford, CT: Lyons, 2013), 126, 241.

44. Moses, *Public Works*, 24, 458, 573, 757.

45. Robert Alden, "The Fair Resumes Today with Many New Exhibits," *New York Times*, April 21, 1965.

46. Philip Benjamin, "Humphrey Stars as Show Reopens," *New York Times*, April 22, 1965.

47. "Vice President's Pause for Prayer," National Catholic Welfare Council News Service (Domestic), April, 27, 1965, Msgr. James Asip Collection, Vatican Pavilion, box 1, RCDBA.

48. "US to Intensify Sea Surveillance," *New York Times*, April 21, 1965.

9. The Vatican Pavilion

1. Jitka Jonová, "Heritage Preservation and Sacred Art after the Second Vatican Council," *Acta Universitatis Carolinae Theologica* 7, no. 1 (2017): 193.

2. Jonová, 193.

3. Jonová, 200.

4. Walter M. Abbott, ed., *The Documents of Vatican II: All Sixteen Official Texts Promulgated by the Ecumenical Council, 1963–1965* (New York: Herder and Herder, 1966), 174–77.

5. Flynn, "Vatican Pavilion," 11.

6. Permission to exceed 120 feet was allowed only for exhibits of the host country, state, or city. Thornton to Moses, September 11, 1962, box 286, NYPL-NYWF.

7. R. T. Bozak, "The Vatican Pavilion," *Audio*, April 1965, 24.

8. Status Report on the carillon of bells, September 7, 1963, box SC93, folder 15, AANY.

9. John Klein was also the official carillonneur at the Seattle World's Fair (1962) and at the Brussels World's Fair (1958), where he played daily concerts at the Vatican pavilion.

10. The Coca-Cola carillon played alternate-hour programmed recitals that lasted about five minutes. "Coke originally wanted their carillon to play much longer but were persuaded that it might be viewed by some as 'too much of a good thing.'" From the unpublished manuscript, "Schulmerich Electronic Carillons New York World's Fair 1964–1965," provided by Schulmerich Carillons, a Division of the Verdin Company, Cincinnati.

11. Rockefellers to Cardinal Spellman, dated July 12, 1962, box SC94, folder 3, AANY.

12. Margaret Delaney, "Former Eurekan Helps Design Symbolic Panels for World's Fair," *Eureka (CA) Humboldt Standard*, April 23, 1964.

13. St. Anthony's Monastery, Kennebunkport, Maine. Interior decorations of the monastery's chapel were designed, produced, and arranged by Jonynas.

14. Bozak, "Vatican Pavilion," 24.

15. Bozak, 24.

16. Larry Zide, "Inzide Audio," *Audio*, April 1965, 36.

17. Unless otherwise noted, exhibit descriptions are from a listing of exhibits from the Msgr. James Asip Collection, Vatican Pavilion, box 1, folder 1, RCDBA.

18. "The Bible in Bronzes: Stanley Bleifeld, a Young Sculptor, Supplies Some Refreshing Surprises," *Life*, June 28, 1963, 105–7.

19. "Prophecies in Clay," reprint, Msgr. James Asip Collection, Vatican Pavilion, box 1, folder 1, RCDBA.

20. Justin Zhuang, "Design History 101: Quietly Beautiful Work by the Illustrator Who Drew the Four Seasons Logo," April 9, 2015, https://eyeondesign.aiga.org/design-history-101-quietly-beautiful-work-by-illustrator-who-drew-four-seasons-logo/.

21. Description of some of the exhibits and art items, box SC92, folder 6, AANY.

22. Reprint from the *Providence Visitor*, January 10, 1964, Msgr. James Asip Collection, Vatican Pavilion, box 1, folder 1, RCDBA.

23. Naj Wintoff, "On the Scene: Duval Opens 'If Not Now, When' Exhibit at Keene Arts," *Lake Placid (NY) News*, August 12, 2021, https://www.lakeplacidnews.com/opinion/columns/2021/08/12/on-the-scene-duval-opens-if-not-now-when-exhibit-at-keene-arts/.

24. Wintoff, "On the Scene."

25. Description of pavilion art displays, Msgr. James Asip Collection, Vatican Pavilion, box 1, RCDBA.

26. Description of pavilion art displays, Msgr. James Asip Collection, Vatican Pavilion, box 1, RCDBA.

27. Archdiocese of New York, box SC92, folder 6, AANY. After the fair, the United Church of Christ acquired the banner for its headquarters in Cleveland. A copy of the banner is in a private collection, and another is owned by the Corita Art Center, by way of Fullerton Junior College in California. From Rose Pacatte, *Corita Kent: Gentle Revolutionary of the Heart* (Collegeville, MN: Liturgical, 2017), 29.

28. Susan Dackerman, "Corita Kent and the Language of Pop," in *Corita Kent and the Language of Pop*, ed. Susan Dackerman (Boston: Harvard Art Museums, 2015), 15–16.

29. Pacatte, *Corita Kent*, 29.

30. Pacatte, 1.

31. Pacatte, 33.

32. Dackerman, "Corita Kent and the Language of Pop," 27.

33. Corita Kent, interview, April 6, 1976, 10, https://static.library.ucla.edu/oralhistory/pdf/masters/21198-zz0008z9kb-5-master.pdf.

34. Kent interview, 37. This referenced work was *Christ and Mary*, 1954.

35. "All You Need Is Love: Pictures, Words and Worship," Eye Magazine, Spring 2000, eyemagazine.com/feature/article/all-you-need-is-love-pictures-words-and-worship.

36. Artists Rights Society, Corita Kent x Chloé SS21 Collection: "A Season in Hope," October 1, 2020, https://arsny.com/blog/corita-kent-chloe-ss21-collection-a-season-in-hope/.

37. *1965 Official Guide*, 6.

38. Luke 15:4–6, John 10:11, Psalm 22:1–2, Isaiah 40:11.

39. Pavilion details, Msgr. James Asip Collection, Vatican Pavilion, box 1, folder 1, RCDBA. An additional note to the Duval windows was Mrs. Duval's role in preparing commentary for the clergy and parishioners on her husband's interpretations of subject matter and aesthetic goals for each commission.

40. Meeting memo, Kiff Voss and Franken architects, September 12, 1963, box 70, folder 6, BRTD-NYPLPA.

41. Directors' meeting minutes, January 22 and 29, 1965, Office of the Bishop, Vatican Pavilion, box 1, folder 1, RCDBA.

42. Directors' meeting minutes, January 22 and 29, 1965.

43. Helen Bradfield, Joan Pringle, and Judy Ridout, *Art of the Spirit: Contemporary Canadian Fabric Art* (Toronto: Dundurn, 1992), 10.

44. My gratitude to Father James Flint, OSB, for his explanation of this centuries-old calendar determination.

45. Vatican Pavilion press release AR1–2300, "Original Liturgical Banners Hang in Vatican Pavilion," Msgr. James Asip Collection, Vatican Pavilion, box 1, folder 1, RCDBA.

46. Thomas D. O'Connor, letter to the editor, *Homiletic and Pastoral Review* 96, no. 2 (November 1965): 8.

47. Fragments of Caius, "I.—From a Dialogue or Disputation against Proclus," from the Ante-Nicene Fathers, vol. 5, ccel.org/ccel/schaff/anf05.v.iii.html.

48. Philip H. Dougherty, "Church Pavilions Use Volunteers," *New York Times*, July 20, 1964, 36.

49. Melissa Renn, "Within Their Walls: LIFE Magazine's 'Illuminations,'" *Archives of American Art Journal* 53, nos. 1/2 (Spring 2014): 33.

50. Spellman Museum of Stamps and Postal History, Spellmanmuseum.org/history.html. Cardinal Spellman's stamp collection now forms, in part, the holdings of the Cardinal Spellman Museum of Stamps and Postal History. The museum is a not-for-profit, 501c3, independent and self-supporting institution.

51. Giuseppe Barbieri, "An American Artist in Italy: 1950–1962," in *William G. Congdon: An American Artist in Italy*, William G. Congdon Foundation (Vicenza, Italy: Terra Ferma, 2001), 36–37.

52. William G. Congdon, *In My Disc of Gold* (New York: Reynal, 1962), 24.

53. Exhibit of "Message of Peace" (ten large oil paintings) by Outstanding Mexican Artist-Muralist Andrés Salgó, box SC94, folder 7, AANY.

54. Gregory Smith, "The Church at the Fair," *Scapular*, March–April 1964, 4–5.

10. The Agony and the Ecstasy

1. Robert Moses to Cardinal Spellman, undated telegram; Thomas Deegan to Robert Moses, January 22, 1965, NYPL-NYWF.

2. Georgina Louise White, *The Pieta in the Night Sky*, 2007 New York University blog post (no longer accessible).

3. Robert C. Doty, "Pope Hails 'Voice' of Michelangelo," *New York Times*, April 19, 1964.

4. Robert Alden, " 'Pieta' Unveiled in Glow of Blue," *New York Times*, April 20, 1964, 1.

5. Alden, 1.

6. Alden, 2.

7. John Canaday, "Happy New Year! Thoughts on Critics and Certain Painters as the Season Opens," *New York Times*, September 6, 1959.

8. John Canaday, "Setting for the 'Pieta': A Critique," *New York Times*, April 20, 1964, 20. Mielziner later learned that on the opening day, floodlights placed in front of the Plexiglas protective screens to light the dignitaries at the opening ceremony offset the subtle lighting plan. Mielziner to Gorman, February 10, 1965, box SC92, folder 2, AANY.

9. John Walker to Cardinal Spellman, April 28, 1964, box 94, folder 7, AANY.

10. John Canaday, "The Pieta and an Avalokiteshvara," *New York Times*, July 26, 1964, 8x.

11. Msgr. Timothy J. Flynn to Mielziner, September 22, 1964, box 16, BRTD-NYPLPA.

12. Mielziner to Msgr. Timothy J. Flynn, July 27, 1964, box 16, BRTD-NYPLPA.

13. Mielziner to Rev. Father Theodore, Holy Spirit Monastery, November 4, 1964, box 16, BRTD-NYPLPA.

14. Arthur Miller to Mielziner, May 9, 1964, box 16, BRTD-NYPLPA.

15. Kinney, *Saga of a Statue*, 48.

16. *1965 Official Guide*, 1965.

17. Robert Bosc, "A European Looks at the New York World's Fair," *America*, July 17, 1965, 77.

18. "Vatican Pavilion's Knight Volunteers," *Columbia*, November 1964, 17–19.

19. Father John J. Gorman to Robert Moses, May 5, 1965, Office of the Bishop, Vatican Pavilion 1965, box 1, RCDBA.

20. "They Know the Answers," *Catholic Advocate*, June 18, 1964.

21. Press release, box SC94, folder 12, AANY.

22. Lauren G. Ziarko, archivist, Manhattanville College, email correspondence with author, January 4, 2019.

23. Mielziner to Kinney, November 5, 1963, box 70, BRTD-NYPLPA.

24. Bozak, "Vatican Pavilion," 24.

25. Mielziner to Msgr. Gorman, September 1, 1964, box 70, BRTD-NYPLPA.

26. Msgr. Gorman to Mielziner, September 5, 1964, box 70, BRTD-NYPLPA.

27. Paul Trautvetter to Msgr. Gorman, June 24, 1964, box 70, BRTD-NYPLPA.

28. "Pieta to Spend Winter in Vatican Pavilion," *Tablet*, October 15, 1964, Msgr. James Asip Collection, Vatican Pavilion, box 1, folder 1, RCDBA.

29. Kinney to the Members of the Vatican Pavilion Committee, October 28, 1964, box SC91, folder 10, AANY.

30. *Tablet*, October 15, 1964, Msgr. James Asip Collection, Vatican Pavilion, box 1, RCDBA.

31. Bob Considine, "On the Line: People . . . Places . . . Pieta," *New York Journal*, February 3, 1965, Msgr. James Asip Collection, Vatican Pavilion, box 1, RCDBA.

32. Rev. Asip to Bishop McEntegart, February 7, 1963, Msgr. James Asip Collection, Vatican Pavilion, box 1, RCDBA.

33. "Visiting Princesses Spend Day Visiting," *New York Times*, May 8, 1965, 19.

34. "Pope Paul Donates His Jeweled Tiara to the Poor of the World," *New York Times*, November 14, 1964.

35. "Papal Tiara Made Gift to U.S. Catholics," Catholic News Service—Newsfeeds, December 1, 1964, 3–4.

36. Moses to Spellman, December 12, 1964, box 286, NYPL-NYWF.

37. "The Pope's Tiara," *Catholic Digest*, April 1965, 60–61. Following the tiara's display at the Vatican pavilion and at St. Patrick's Cathedral, Monsignor Nolan said he "decided to use the tiara as bait to attract people who might then pay some attention to what his office does: namely, provide help for the poor, the sick, and refugees in eighteen countries of the Near East, including Egypt, Bulgaria, Greece, Iran, Lebanon, India, Turkey, and Palestine. The tiara was shown at the National Catholic Educational Association convention, where twenty thousand people came to see the exhibit there. Similarly, the tiara and accompanying Catholic Near East Welfare Association display traveled to Knights of Columbus and Daughters of Isabella conventions in Miami, to a Catholic Students' Mission Crusade convention at Notre Dame, to individual churches and cathedrals, and to the Liturgical Conference convention in Kansas City. A full-page ad in the *New York Times* (paid for by Macy's) publicized the pope's concern for the poor. Later the tiara exhibit went to Macy's in New Haven, Connecticut, and to a shopping center outside Albany, to a bank in Philadelphia, and to a bank in Trenton, New Jersey, and more requests started coming in: from the Gimbels department stores in Philadelphia and Milwaukee and from department stores in Pittsburgh and San Antonio. Tens of thousands of dollars were donated on this "tour." See "Pope Paul's Tiara Raises a New $64,000 Question," *Catholic Transcript*, November 1967.

38. "Vatican Pavilion Plans to Show Disputed Statue by Michelangelo," *New York Times*, April 3, 1965, 27.

39. National Gallery of Art, "Giovanni Francesco Susini," nga.gov/collection/art-object-page.133635.html#overview.

40. The original idea was to exhibit a garden of peas to replicate more faithfully Mendel's garden, but David Burpee of Burpee Seeds advised against it, as pea blossoms would not last into the summer. David Burpee to William Laurence, April 12, 1965, box 286, NYPL-NYWF.

41. Press release, June 17, 1965, box 286, NYPL-NYWF; Philip H. Dougherty, "Mendel Garden Is Opened at Fair," *New York Times*, June 21, 1965, 34.

42. Stuart Constable, vice president, operations, to all exhibitors and concessionaires, March 16, 1965, box 286, NYPL-NYWF.

43. Kinney to John Young, June 11, 1965, box SC91, folder 10, AANY.

44. Howard Vogel to Milton P. Kayle, February 25, 1964, box 286, NYPL-NYWF.

45. Five different designs were made by Th. Marthinsen of Norwegian Silversmiths: the *Pietà*, two of the *Good Shepherd*, the pavilion chapel, crown of the chapel, and papal coat of

arms. Three of the designs were in sterling silver and three in silver plate, on four-inch demitasse spoons. Msgr. James Asip Collection, Vatican Pavilion, box 1, folder 1, RCDBA.

46. Msgr. Asip to Bishop McEntegart, May 3, 1963, Msgr. James Asip Collection, Vatican Pavilion, box 1, folder 1, RCDBA.

47. Meeting minutes, July 1, 1963, box SC93, folder 3, AANY.

11. City's Embrace

1. Aldo Cortesis, "Vatican Hints Role in 1964 Fair Here," *New York Times*, September 4, 1960, 1.

2. John Cooney, *The American Pope: The Life and Times of Francis Cardinal Sepllman* (New York: Times Books, 1984), 215.

3. John Young to Robert Moses, press release, December 6, 1963, box 286, NYPL-NYWF.

4. Tom Deegan to Robert Moses, December 24, 1966, box 286, NYPL-NYWF.

5. Moses to Pope Paul VI, December 1964, box 287, NYPL-NYWF.

6. "Fair Again Asks Pope to Attend," *New York Times*, December 31, 1964, 20.

7. "Vatican Indicates Pope Won't Come to the Fair," *New York Times*, January 3, 1965.

8. John Young to Robert Moses, memorandum, June 25, 1965, box 287, NYPL-NYWF.

9. E. Vagnozzi to Robert Moses, August 30, 1965, box 287, NYPL-NYWF.

10. Peter Hebblethwaite, *Paul VI: The First Modern Pope* (New York: Paulist, 1993), 459.

11. Hebblethwaite, 435.

12. Homer Bigart, "An Historic Mission," in *The Pope's Journey to the United States*, ed. A. M. Rosenthal and Arthur Gelb (New York: Bantam Books, 1965) (hereafter *Pope's Journey*), 1.

13. Abe Rosenthal, introduction to *Pope's Journey*, ii.

14. David W. Dunlap, "Here's the Pope. Where's the Paper?," *New York Times*, July 10, 2015.

15. On Pope Paul VI's 1970 visit to the Philippines, Marcinkus tackled a man armed with a knife who lunged at the pope at the Manila Airport. In the 1980s Marcinkus, as head of the Vatican Bank, was implicated in the collapse of Banco Ambrosiano; he maintained his innocence but was relieved of his position. He retired to Sun City, Arizona, were he died in 2006. Ron Grossman, "Archbishop Paul Marcinkus 1922–2006," *Chicago Tribune*, June 22, 2006, https://www.chicagotribune.com/news/ct-xpm-2006-02-22-0602220046-story.html.

16. Leo McFadde, "Papa Travels Timed to the Minute," *Catholic Advocate*, August 6, 1970, 2.

17. McFadde, 2.

18. Sydney H. Schanberg, "Sharing the Financial Burden," in *Pope's Journey*, 76.

19. Bernard Weinraub, "A Transfixed City," in *Pope's Journey*, 52.

20. Martin Arnold, "It's Almost as If They Were All in Church," in *Pope's Journey*, 58.

21. Peter Kihss, "A Formidable Day for the Police," in *Pope's Journey*, 64.

22. Peter Grose, "Problems of Protocol," in *Pope's Journey*, 74–75.

23. Kihss, "Formidable Day for the Police," 64.

24. Theodore Jones, "In the Streets of Harlem," in *Pope's Journey*, 46–47.

25. Bigart, "Historic Mission," 4.

26. Arnold, "It's Almost as If They Were All in Church," 58.

27. National Catholic Welfare Conference News Service, October 5, 1965, 10.

28. Arnold, "It's Almost as If They Were All in Church," 56.

29. Kihss, "Formidable Day for the Police," 61.

30. Murray Schumach, "From Sleaziest Slum to Sleekest Luxury," in *Pope's Journey*, 43.

31. Hebblethwaite, *Paul VI*, 435.

32. Bigart, "Historic Mission," 5.

33. Bundy also advised Johnson on how to proceed if the pope were to raise the issue of Cardinal Mindszenty, whom Bundy wanted removed from the American Embassy in Budapest. Only the pope could initiate this request, and Bundy acknowledged it would be difficult even for him. As Mindszenty remained at the legation until 1971, it would seem that this topic, had it been raised, was not acted upon. See McGeorge Bundy, "No. 302, Memorandum from the President's Special Assistant for National Security Affairs (Bundy) to President Johnson, Washington, October 3, 1965," In *Foreign Relations of the United States, 1964–1968, Volume XII, Western Europe.*

34. Tom Wicker, "A Talk with the President," in *Pope's Journey*, 14–21.

35. Michael G. Kort, *The Reexamination of the Vietnam War* (Cambridge: Cambridge University Press, 2017), 117.

36. Bundy, "No. 302."

37. Roy Palmer Domenico, "America, the Holy See and the War in Vietnam," in *Papal Diplomacy in the Modern Age*, ed. Peter C. Kent and John F. Pollard (Westport, CT: Praeger, 1994), 209.

38. Wicker, "Talk with the President," 16.

39. Luci Johnson's baptism into the Catholic Church was criticized in some Catholic circles, since she had already been baptized in the Episcopalian Church. Father Thomas Stansky of Milwaukee, a spokesman for the Vatican's Secretariat for Promoting Christian Unity, called this act "regrettable," as baptism in the Episcopalian Church was deemed "valid." It was a reflection of the lingering distrust between Catholics and Protestants. "Vatican Calls It Regrettable: Conversion Okay, but Luci's Baptism Blasted," *Desert Sun* (Palm Springs, CA), July 6, 1965.

40. Wicker, "Talk with the President," 20.

41. Kihss, "Formidable Day for the Police," 64.

42. Hebblethwaite, *Paul VI*, 439.

43. Drew Middleton, "The Pope's Plea: 'War Never Again!,'" in *Pope's Journey*, 21.

44. Raymond Daniell, "A Word for Everyone," in *Pope's Journey*, 27.

45. A. M. Rosenthal, "Pope's Visit to New York and Peace Appeal at U.N.," *New York Times*, October 11, 1965, 59.

46. Hebblethwaite, *Paul VI*, 441.

47. Daniell, "Word for Everyone," 28. In less than four minutes and with only five bidders, the cross and ring presented by Pope Paul VI to the United Nations were sold at auction for $64,000. John L. Marion of the Parke-Bernet Galleries opened the public auction by reading statements from U Thant and Alberto Giovannetti, permanent observer of the Holy See to the UN. Marion concluded the statements and the ground rules by saying "this is probably the shortest and most selective auction in history." Harry Levinson, president of Levinson's Jewels in Chicago, made the winning bid. It was decided by U Thant, in consultation with all concerned, that the proceeds of the sale would be distributed in equal parts to the UN Children's Fund, the Office of the UN High Commissioner for Refugees, the UN Relief and Works Agency for Palestine Refugees in the Near East, and the Freedom from Hunger Campaign. When asked what his plans were for the papal cross and ring, Levinson said, "I would like to see them in a museum," and that he would try to sell them intact to a purchaser who would in turn donate them to a museum. If not, they would eventually be broken up (Catholic News Service, Newsfeeds, November 3, 1967).

48. Visit of His Holiness Pope Paul VI to the United Nations, https://www.unmultimedia.org/s/photo/detail/369/0369185.html.

49. George Dugan, "The Other Faiths," in *Pope's Journey*, 29–32.

50. Dugan, 29.

51. John Canaday, "The Dilemma of Church Art," in *Pope's Journey*, 86.

52. Kihss, "Formidable Day for the Police," 64; National Catholic Welfare Conference News Service, October 5, 1965, 9.

53. William E. Farrell, "90,000 Amens," in *Pope's Journey*, 34.

54. Farrell, 34.

55. Gerald Moore, *LIFE Story: The Education of an American Journalist* (Albuquerque: University of Arizona Press, 2016), 192.

56. Moore, 195. Bill Eppridge is best remembered for his iconic photo of Senator Robert F. Kennedy mortally wounded on the floor of the Los Angeles Ambassador Hotel.

57. Kihss, "Formidable Day for the Police," 62.

58. Pope Paul statements at Yankee Stadium, in *Pope's Journey*, 115–17.

59. Kihss, "Formidable Day for the Police," 60.

60. Monsignor Paul Marcinkus to Bishop-elect Terrence Cooke, September 22, 1965, box 287, NYPL-NYWF.

61. As told to Kinney by Cardinal Spellman, in Kinney, *Saga of a Statue*, 73.

62. Kihss, "Formidable Day for the Police," 60.

63. John Cogley, "The Return to Rome," in *Pope's Journey*, 40.

64. Cogley, 40.

65. Cogley, 41.

66. Cogley, 41.

67. John C. Bennett, "Pope Paul's Visit," *Christianity and Crisis*, November 1, 1965, 1.

68. Rev. James Asip to Dioceses, April 1, 1966, Msgr. James Asip Collection, Vatican Pavilion, box 1, folder 1, RCDBA.

69. Bishop McEntegart to Rev. James Asip, April 2, 1966, Msgr. James Asip Collection, Vatican Pavilion, box 1, folder 1, RCDBA; Cardinal Spellman to Cardinal Cushing, January 28, 1966, SC94-F10, AANY.

70. Vatican Pavilion (New York World's Fair) Inc., Attendance Highlights, 1964–1965, SC94-F12, AANY.

71. Handwritten letter to Robert Moses, undated, box 286, NYPL-NYWF.

12. *Ciao*, New York

1. Ed Wilkinson, "Readers Recall World's Fair Memories," *Tablet*, September 3, 2014, https://thetablet.org/readers-recall-worlds-fair-memories/.

2. Wilkinson, "Readers Recall World's Fair Memories."

3. Jim Draeger and Daina Penkiunas, "The Space Age Journey of Wisconsin's 1964 World's Fair Pavilion," *Wisconsin Magazine of History* 97, no. 4 (Summer 2014): 24.

4. Edward M. Kinney, "The Vatican Pavilion after the Fair," *Catholic Market*, July–August 1965, 70.

5. Pacatte, *Corita Kent*, 29.

6. My gratitude to Bill Cotter for the location of this piece.

7. "Vatican Fair Banners Sold," *Catholic Transcript*, August 19, 1966.

8. Kinney, "Vatican Pavilion after the Fair," 70.

9. Ed Wilkinson, "50 Years Later, World's Fair Nostalgia Still Alive," *Tablet*, August 20, 2014, https://thetablet.org/50-years-later-worlds-fair-nostalgia-still-alive/. The pedestal appears to have been cut down to accommodate the smaller chapel space. The second *Pietà* replica was placed at St. Joseph's Seminary at Dunwoodie, in Yonkers, New York. Edward Kinney relayed that the two marbles were shipped to New York from Livorno, Italy, and upon arrival were inspected by the three members of the transport committee, Kearns, McNally, and

Kinney. Both arrived intact, but it was discovered that one of the inner crates had a nail protruding about three-quarters of an inch, which resulted in a sizable chip off the marble statue.

10. Secretary of state of São Paulo state to Robert Moses, July 28, 1965, box 287, NYPL-NYWF.

11. Much of this chapter is based on Edward M. Kinney's *Saga of a Statue*, his self-published account of the transport of the *Pietà*.

12. Kinney, *Saga of a Statue*, 76.

13. Kinney, 78.

14. Edward M. Kinney to Members of the Vatican Pavilion Committee, October 18, 1965, box SC92, folder 10, AANY.

15. Kinney, *Saga of a Statue*, 79.

16. Kinney, 80.

17. Kinney, 80.

18. Kinney, 82.

19. Kinney, 82.

20. Kinney, 84.

21. John Murray obituary, Fordham News, October 2, 2013, https://news.fordham.edu/arts-and-culture/john-murray-fcrh-57-who-brought-michelangelos-pieta-to-1964-worlds-fair-dies-at-83/.

22. "Vatican Forbids Art Loans Abroad," *New York Times*, September 10, 1965, 40.

23. Walter Persegari, "The Vatican Collections: The Papacy and Art," *Museum International* 38, no. 4 (January/December 1986): 213.

24. John Foley, "If New Church in Groton Looks Familiar, It Should," *Catholic Transcript*, November 17, 1967, 9.

25. Foley, "If New Church in Groton Looks Familiar, It Should."

26. Courtesy of St. Mary, Mother of the Redeemer Catholic Church, Groton, CT, church history booklet, undated, 2.

27. St. Mary, Mother of the Redeemer Catholic Church, http://www.stmarysgroton.org/about-us/our-history/.

28. Foley, "If New Church in Groton Looks Familiar, It Should."

29. Foley, "If New Church in Groton Looks Familiar, It Should."

30. Robert Moses to John J. Gorman, May 6, 1966, box 287, NYPL-NYWF.

31. WNYC, "Flushing Meadows Corona Park, Queens Dedication Ceremonies," NYPR (New York Public Radio) Archive Collections, wnyc.org/story/flushing-meadows-corona-park-queens-dedication-ceremonies/.

32. Murray Schumach, "Moses Gives City Fair Site as Park," *New York Times*, June 4, 1967, 1.

33. John R. Lewis, *Discover Flushing Meadows–Corona Park: A Guidebook* (New York: n.p.), 15.

34. Joseph P. Laycock, *The Seer of Bayside: Veronica Lueken and the Struggle to Define Catholicism* (New York: Oxford University Press, 2015), 6.

Epilogue

1. Gay Talese, *The Kingdom and the Power* (New York: World, 1969), 102.

2. Talese, 101.

3. Russell Shaw, *Catholics in America: Religious Identity and Cultural Assimilation from John Carroll to Flannery O'Connor* (San Francisco: Ignatius, 2016), 95.

4. "Spellman Death Marks End of Era," *National Catholic Reporter* 4, no. 8 (December 1967): 1.

5. Thomas J. Shelley, "Francis J. Spellman: The Evolution of an Ecclesiastical Career," *American Catholic Studies* 114, no. 3 (Fall 2003): 92.

6. Ballon and Jackson, *Robert Moses and the Modern City*, 199–200.

7. Michael Powell, "In Queens, an Oasis Neglected by the City," *New York Times*, October 2, 2012, 24.

8. Paul Hofmann, "Vatican Expert Foresees a 'Perfect Restoration of the Pieta,'" *New York Times*, May 23, 1972, 5.

9. As told to French journalist Yves Stavrides, in Crista Kramer von Reisswitz, "Did Paul VI Want to Sell the Pieta?," *Inside the Vatican*, March 2002, 11.

10. Reisswitz, "Did Paul VI Want to Sell the Pieta?"

Bibliography

I began my research in the New York Public Library Archives, the official depository for documents of the World's Fair Corporation, whose records are organized by pavilion. As I read through the files, I was struck by the enthusiasm of Robert Moses to enlist the Vatican to go along with his plan to obtain the *Pietà* for the World's Fair. The humble and deferential tone of his correspondence with Cardinal Spellman contrasted sharply with his abrasive public persona. The library's Billy Rose Theatre Division, the most comprehensive archives devoted to the theatrical arts, housed the papers of Joseph "Jo" Mielziner, the Broadway scenic designer responsible for staging the *Pietà* at the fair. His commitment to creating a setting worthy of the masterpiece and the technical challenges involved in accomplishing the task fascinated me.

Because the construction and the operation of the Vatican pavilion were a collaboration between the Archdiocese of New York and the Diocese of Brooklyn, the archives of both were critical to tracking how the *Pietà* made its way to America. The archives of the Archdiocese of New

York, kept at St. Joseph's Seminary in Dunwoodie, and the Archives of the Diocese of Brooklyn provided rich resources on the development of the pavilion and the planning of the programming and installation there of Christian artwork from the early church, as well as on the tradition-shattering liturgical shifts of Vatican II. Having grown up in the Chicago area, I did not know who Francis Cardinal Spellman was before I began this project. After days of combing through volumes of his correspondence, I came to understand the far-reaching scope of his influence, all but lost and forgotten in these more secular times.

Abbot, Walter M., ed. *The Documents of Vatican II: All Sixteen Official Texts Promulgated by the Ecumenical Council, 1963–1965*. New York: Herder and Herder, 1966.

Arnold, Martin. "It's Almost as If They Were All in Church." In Rosenthal and Gelb, *Pope's Journey*, 55–60.

Artwell, Henry. "'I Looked at It with Awe'—as Thousands Do Daily." *Catholic Courier*, May 7, 1964.

Ault, Julie, and Martin Beck. "All You Need Is Love: Pictures, Words and Worship." *Eye*, Spring 2020. eyemagazine.com/feature/article/all-you-need-is-love-pictures-words-and-worship.

Ballon, Hilary. "Lincoln Square Title I." In Ballon and Jackson, *Robert Moses and the Modern City*, 279–82.

Ballon, Hilary. "Robert Moses and Urban Renewal: The Title I Program." In Ballon and Jackson, *Robert Moses and the Modern City*, 94–115.

Ballon, Hilary, and Kenneth T. Jackson, eds. *Robert Moses and the Modern City: The Transformation of New York*. New York: W. W. Norton, 2008.

Barbieri, Giuseppe. "An American Artist in Italy: 1950–1962." In *William G. Congdon: An American Artist in Italy*, 30–37. William G. Congdon Foundation. Vicenza, Italy: Terra Ferma, 2001.

Bennett, John C. "Pope Paul's Visit." *Christianity and Crisis*, November 1, 1965, 1–2.

Bigart, Homer. "An Historic Mission." In Rosenthal and Gelb, *Pope's Journey*, 1–7.

Bogart, Michele H. *The Public Art of Robert Moses*. National Trust for Historic Preservation, September 25, 2018. https://savingplaces.org/stories/the-public-art-of-robert-mosess-new-york.

Bogart, Michele H. *Sculpture in Gotham: Art and Urban Renewal in New York City*. London: Reaktion Books, 2018.

Bosc, Robert. "A European Looks at the New York World's Fair." *America*, July 17, 1965.

Bozak, R. T. "The Vatican Pavilion." *Audio*, April 24, 1965.

Bradfield, Helen, Joan Pringle, and Judy Ridout. *Art of the Spirit: Contemporary Canadian Fabric Art*. Toronto: Dundurn, 1992.

Brandt, Kathleen Weil-Garris. "Michelangelo's Pietà for the Cappella del Re di Francia." In *Michelangelo: Selected Scholarship in English (Book 1)*, edited by William E. Wallace, 77–117. New York: Routledge, 1996.

Brooks, John. "Diplomacy at Flushing Meadow." Onward and Upward with the Arts, *New Yorker*, June 2, 1963.

Bundy, McGeorge. "No. 302. Memorandum from the President's Special Assistant for National Security Affairs (Bundy) to President Johnson." October 3, 1965. In *Foreign Relations of the United States, 1964–1968, Volume XII, Western Europe*. https://history.state.gov/historicaldocuments/frus1964-68v12/d302.

Canaday, John. "The Dilemma of Church Art." In Rosenthal and Gelb, *Pope's Journey*, 85–87.

Canaday, John. *Embattled Critic: Views on Modern Art*. New York: Farrar, Straus and Giroux, 1965.

Caro, Robert. *The Power Broker: Robert Moses and the Fall of New York*. New York: Knopf, 1974.

Carrillo, Elisa A. "The Italian Church and Communism, 1943–1963." *Catholic Historical Review* 77, no. 4 (1991): 644–57.

Catholic Courier Journal. "X-rays Show Repairs to Broken Pieta." April 23, 1964.

Catholic Digest. "The Pope's Tiara." April 1965, 60–61.

Catholic Digest. "He Sets the Stage." July 1962.

Catholic News Service. "Packing the Pieta Revisited: American Know-How Paid Off." Newsfeeds, June 2, 1972.

Catholic News Service "Papal Tiara Made Gift to U.S. Catholics." Newsfeeds, December 1, 1964, 3–4.

Catholic Standard and Times. "Exclusion Would Be Discrimination." January 3, 1958.

Catholic Standard and Times. "Urban Renewal Unit Formed in New York." May 19, 1961, 13.

Christin, Pierre, and Olivier Balez. *Robert Moses: The Master Builder of New York City*. London: Nobrow, 2014.

Cogley, John. "The Return to Rome." In Rosenthal and Gelb, Pope's Journey, 40–42.

Columbia. "Vatican Pavilion's Knight Volunteers." Vol. 44 (November 1964).

Congdon, William G. *In My Disc of Gold*. New York: Reynal, 1962.

Cooney, John. *The American Pope: The Life and Times of Francis Cardinal Sepllman*. New York: Times Books, 1984.

Corpo, Ryan Di. "The Story behind the Lost Neighborhood Where 'West Side Story' Is Set." *America: The Jesuit Review*, February 25, 2020. https://www.americamagazine.org/arts-culture/2020/02/25/story-behind-lost-neighborhood-where-west-side-story-set.

Cranz, Galen. *The Politics of Park Design: A History of Urban Parks in America*. Cambridge, MA: MIT Press, 1982.

Craven, Thomas. *Men of Art*. New York: Simon & Schuster, 1940.

Craven, Wayne. *Sculpture in America*. London: Cornwall Books, 1984.

Crosby, Donald F. *God, Church, and Flag: Senator Joseph R. McCarthy and the Catholic Church, 1950–1957*. Chapel Hill: University of North Carolina Press, 1978.

Dackerman, Susan. "Corita Kent and the Language of Pop." In *Corita Kent and the Language of Pop*, 15–33. Cambridge, MA: Harvard Art Museums, 2015.

Daniell, Raymond. "A Word for Everyone." In Rosenthal and Gelb, *Pope's Journey*, 26–28.

Davis, Margaret Leslie. *Mona Lisa in Camelot: How Jacqueline Kennedy and Da Vinci's Masterpiece Charmed and Captivated a Nation*. New York: Da Capo, 2008.

Delaney, Margaret. "Former Eurekan Helps Design Symbolic Panels for World's Fair." *Eureka (CA) Humboldt Standard*, April 23, 1964.

Denver Catholic Register. "Staging the Pieta." August 25, 1963.

Domenico, Roy Palmer. "America, the Holy See and the War in Vietnam." In *Papal Diplomacy in the Modern Age*, edited by Peter C. Kent and John F. Pollard, 203–19. Westport, CT: Praeger, 1994.

Draeger, Jim, and Daina Penkiunas. "The Space Age Journey of Wisconsin's 1964 World's Fair Pavilion." *Wisconsin Magazine of History* 97, no. 4 (Summer 2014): 21–60.

Dugan, George. "The Other Faiths." In Rosenthal and Gelb, *Pope's Journey*, 29–32.

Dulles, Avery. *A Testimonial to Grace and Reflections on a Theological Journey*. Kansas City, MO: Sheed & Ward, 1996.

Faggioli, Massimo. *John XXIII: The Medicine of Mercy*. Collegeville, MN: Liturgical, 2014.

Farrell, William E. "90,000 Amens." In Rosenthal and Gelb, *Pope's Journey*, 32–38.

Florman, Samuel C., "Deus ex Machina." *Mechanical Engineering* 134, no. 12 (December 2012).

Flynn, Timothy J. "The Vatican Pavilion." *Catholic Market*, January 1964, 104.

Foley, John. "If New Church in Groton Looks Familiar, It Should." *Catholic Transcript*, November 17, 1967.

Frankfurter, Alfred. "The Michelangelo Scandal." *ARTnews* 61, no. 3 (May 1962).

Frankfurter, Alfred. "On Making the World Safe for Art." *ARTnews* 62, no. 10 (May 1964).

Fredericks, Suzanne P. *Marshall Fredericks, Sculptor*. Saginaw, MI: Saginaw Valley State University, 2003.

Fussiner, Howard. "Art at the Fair." *College Art Journal* 18, no. 1 (1958): 68–71. https://www.jstor.org/stable/i231572.

Gabler, Neal. *An Empire of Their Own: How Jews Invented Hollywood*. New York: Anchor Books, 1988.

Gannon, Robert I. *The Cardinal Spellman Story*. Permabook. New York: Doubleday, 1963.

Gelb, Arthur. *City Room*. New York: Putnam, 2003.

Gordon, James B. "Packing of Michelangelo's 'Pieta.'" *Studies in Conservation* 12, no. 2 (May 1967): 57–69.

Gratz, Roberta Brandes. *The Battle for Gotham: New York in the Shadow of Robert Moses and Jane Jacobs*. New York: Nation Books, 2010.

Grose, Peter. "Problems of Protocol." In Rosenthal and Gelb, *Pope's Journey*, 74–75.

Hammer, Ellen. *A Death in November: America in Vietnam, 1963*. New York: Oxford University Press, 1987.

Hartt, Frederick. *History of Italian Renaissance Art: Painting, Sculpture, Architecture*. New York: Harry N. Abrams, 1979.

Harvard University. *Risk Management & Audit Services*. https://rmas.fad.harvard.edu/pages/goods-equipment-transit-risk-discussion.

Haynes, John E. *Red Scare or Red Menace?* Chicago: Ivan R. Dee, 1996.

Hebblethwaite, Peter. *Paul VI: The First Modern Pope*. New York: Paulist, 1993.

Hebblethwaite, Peter. *Pope John XXIII: Shepherd of the Modern World; The Definitive Biography of Angelo Roncalli*. Garden City, NJ: Image Books, 1987.

Heller, Alfred. *World's Fairs and the End of Progress: An Insider's View*. Corte Madera, CA: World's Fair, 1999.

Henderson, Mary C. *Mielziner: Master of Modern Stage Design*. New York: Backstage Books, 2001.

Higgins, Marguerite. *Our Vietnam Nightmare*. New York: Harper & Row, 1965.

Hoehmann, George A. "'My Eminent Friend of New York': Francis Cardinal Spellman and the Second Vatican Council." MA thesis, St. Joseph's Seminary (Yonkers, NY), 1992.

Jacobs, Seth. *Cold War Mandarin: Ngo Dinh Diem and the Origins of America's War in Vietnam 1950–1953*. Lanham, MD: Rowman & Littlefield, 2006.

Jones, Theodore. "In the Streets of Harlem." In Rosenthal and Gelb, *Pope's Journey*, 45–47.

Jonová, Jitka. "Heritage Preservation and Sacred Art after the Second Vatican Council." *Acta Universitatis Carolinae Theologica* 7, no. 1 (2017): 193–206.

Jurkowlaniec, Grżyna. "A Miracle of Art and Therefore a Miraculous Image: A Neglected Aspect of the Reception of Michelangelo's Vatican Pietà." *Artibus et Historiae* 36, no. 7 (2015).

Katolin, B. "The Servant of Three Gentlemen." *Trud* (Moscow), no. 29, May 31, 1961; translation, March 29, 1981, 5.

Kendrick, Alexander. *The Wound Within: America in the Vietnam Years, 1945–1974*. Boston: Little, Brown, 1974.

Kent, Korita. Interview, April 6, 1976. University of California, Los Angeles. https://static.library.ucla.edu/oralhistory/pdf/masters/21198-zz0008z9kb-5-master.pdf.

Kihss, Peter. "A Formidable Day for the Police." In Rosenthal and Gelb, *Pope's Journey*, 61–65.

Kinney, Edward M. *The Saga of a Statue*. Self-published, 1989.

Kort, Michael G. *The Reexamination of the Vietnam War*. Cambridge: Cambridge University Press, 2017.

Lane, Barbara G. *The Altar and the Altarpiece: Sacramental Themes in Early Netherlandish Painting*. New York: Harper & Row, 1984.

Laycock, Joseph P. *The Seer of Bayside: Veronica Lueken and the Struggle to Define Catholicism*. New York: Oxford University Press, 2015.

Lebovics, Herman. *Mona Lisa's Escort: André Malraux and the Reinvention of French Culture*. Ithaca, NY: Cornell University Press, 1999.

Lewis, John R. *Discover Flushing Meadows–Corona Park: A Guidebook*. New York: n.p., 1976.

Life. "The Bible in Bronzes: Stanley Bleifeld, a Young Sculptor, Supplies Some Refreshing Surprises." June 28, 1963, 105–7.

Lloyd, R. Scott. "Elder Perry: Mormon Pavilion at 1964 World's Fair Had Impact." Church of Jesus Christ of Latter-Day Saints, September 22, 2014. https://www.churchofjesuschrist.org/church/news/elder-perry-mormon-pavilion-at-1964-worlds-fair-had-impact?lang=eng.

Lomask, Milton. "Return to Orthodoxy." *Sign*, January 1955.

Lombardo, Josef Vincent. *Michelangelo: The Pietà and Other Masterpieces*. New York: Pocket Books, 1965.

L'Osservatore Romano. "Pope John XXIII." Homily of the Holy Father, September 6, 2000. https://www.vatican.va/news_services/liturgy/saints/ns_lit_doc_20000903_john-xxiii_en.html.

Macaulay-Lewis, Elizabeth, and Jared Simard. "From Jerash to New York: Columns, Archaeology, and Politics at the 1964–65 World's Fair." *Journal of the Society of Architectural Historians* 74, no. 3 (2015): 343–64.

MacGregor, Morris J. *Steadfast in the Faith*. Washington, DC: Catholic University of America Press, 2005.

Mackie, Gillian Vallance. *Early Christian Chapels in the West: Decoration, Function and Patron*. Toronto: University of Toronto Press, 2003.

Magnan, Dan. "Shea Was His Field of Dreams—Moses Envisioned It as Jewel of Complex." *New York Post*, October 24, 2000.

Maier, Thomas. *When Lions Roar: The Churchills and the Kennedys*. New York: Crown, 2014.

Manning, Martin J. "Fairs! Fairs! Fairs! The United States Information Agency and U.S. Participation at World's Fairs since World War II." *Popular Culture in Libraries* 2, no. 3 (1994): 1–32.

Mayer, Martin. "Ho Hum, Come to the Fair." *Esquire*, October 1963.

McCaffrey, Lawrence. *The Irish Diaspora in America*. Bloomington: Indiana University Press, 1976.

McFadde, Leo. "Papa Travels Timed to the Minute." *Catholic Advocate*, August 6, 1970.

McNaspy, C. J. "World's Fair Preview." *America*, April 11, 1964.

Middleton, Drew. "The Pope's Plea: 'War Never Again!'" In Rosenthal and Gelb, *Pope's Journey*, 20–25.

Mielziner, Jo. *Designing for the Theatre: A Memoir and a Portfolio*. New York: Atheneum, 1965.

Milleker, Elizabeth J. "A Brief History of the Cast Collection at the Metropolitan Museum of Art." Monograph, Metropolitan Museum of Art, n.d.

Miller, Edward. *Misalliance: Ngo Dinh Diem, the United States, and the Fate of South Vietnam*. Cambridge, MA: Harvard University Press, 2013.

Miller, Marc H. "Something for Everyone: Robert Moses and the Fair." In *Remembering the Future: The New York World's Fair from 1939–1964*, edited by Robert Rosenblum, 45–73. New York: Rizzoli International, 1989.

Molella, Arthur P. "The Human Spirit in an Age of Machines: The Pietà and the Computer at the 1964–1965 New York World's Fair." In *World's Fairs in the Cold War: Science, Technology, and the Culture of Progress*, edited by Arthur P. Molella et al., 96–108. Pittsburgh: University of Pittsburgh Press, 2019.

Moore, Gerald. *LIFE Story: The Education of an American Journalist*. Albuquerque: University of Arizona, 2016.

Morgan, Jack Masey, and Conway Lloyd. *Cold War Confrontations: US Exhibitions in the Cultural Cold War*. New York: Lars Müller, 2008.

Morgan, Joseph G. *The Vietnam Lobby: The American Friends of Vietnam, 1955–1975*. Chapel Hill: University of North Carolina Press, 1997.

Morris, Charles R. *American Catholic: The Saints and Sinners Who Built America's Most Powerful Church*. New York: Random House, 1997.

Moses, Robert. "From Dump to Glory." *Saturday Evening Post*, January 15, 1938.

Moses, Robert. *Public Works: A Dangerous Trade*. New York: McGraw-Hill, 1970.

Moses, Robert. *The Saga of Flushing Meadow, the Valley of Ashes*. New York: Triborough Bridge and Tunnel Authority, 1966.

Murphy, Charles J. V. "The Cardinal and the City." *Fortune*, February 1960.

National Catholic Reporter. "Spellman Death Marks End of Era." December 13, 1967.

Nicoletta, Julie. "Selling Spirituality and Spectacle: Religious Pavilions at the New York World's Fair of 1964–1965." *Buildings & Landscapes: Journal of the Vernacular Architecture Forum* 22, no. 2 (2015): 62–88.

1965 Official Guide: New York World's Fair. New York: Time-Life Books, 1965.

Official Guide Book, Vatican Pavilion, New York World's Fair 1964–1965. New York: Vatican Pavilion (New York World's Fair), 1964.

O'Connor, C. M., and D. Thomas. "Letters to the Editor." *Homiletic and Pastoral Review* 66, no. 2 (1965): 8.

Pacatte, Rose. *Corita Kent: Gentle Revolutionary of the Heart*. Collegeville, MN: Liturgical, 2017.

Persegari, Walter. "The Vatican Collections: The Papacy and Art." *Museum International* 38, no. 4 (January/December 1986): 213–29.

Poletti, Charles. "The Reminiscences of Charles Poletti." Vol. 4. Center for Oral History. New York: Columbia University, 1978.

Reisswitz, Crista Kramer von. "Did Paul VI Want to Sell the Pieta?" *Inside the Vatican*, March 2002.

Renn, Melissa. "Within Their Walls: LIFE Magazine's 'Illuminations.'" *Archives of American Art Journal* 53, nos. 1–2 (Spring 2014): 30–51.

Rice, Louise. *The Altars and Altarpieces of New St. Peter's: Outfitting the Basilica, 1621–1666*. Cambridge: Cambridge University Press, 1997.

Rogers, Adam. "Global Ambition." *Smithsonian*, June 2017.

Rosegrant, Robert G. "Packing Problems and Procedures." *Technical Studies in the Field of the Fine Arts* 10, no. 3 (1942): 138–56.

Rosenthal, A. M. Introduction to Rosenthal and Gelb, *Pope's Journey*, ii.

Rosenthal, A. M., and Arthur Gelb. *The Pope's Journey to the United States*. New York: Bantam Books, 1965.

Samuel, Lawrence R. *The End of the Innocence: The 1964–1965 New York World's Fair*. Syracuse, NY: Syracuse University Press, 2010.

Schanberg, Sydney H. "Sharing the Financial Burden." In Rosenthal and Gelb, *Pope's Journey*, 76–78.

Schumach, Murray. "From Sleaziest Slum to Sleekest Luxury." In Rosenthal and Gelb, *Pope's Journey*, 43–44.

Shaw, Geoffrey D. T. *The Lost Mandate of Heaven: The American Betrayal of Ngo Dinh Diem, President of Vietnam*. San Francisco: Ignatius, 2015.

Shaw, Russell. *Catholics in America: Religious Identity and Cultural Assimilation from John Carroll to Flannery O'Connor*. San Francisco: Ignatius, 2016.

Shelley, Thomas J. *The Bicentennial History of the Archdiocese of New York, 1808–2008*. Strasbourg: Éditions du Signe, 2007.

Shelley, Thomas J. "The Evolution of an Ecclesiastical Career." *American Catholic Studies* 114, no. 3 (2003): 87–92.

Shelley, Thomas J. "Slouching toward the Center: Cardinal Francis Spellman, Archbishop Paul J. Hallinan and American Catholicism in the 1960s." *U.S. Catholic Historian* 17, no. 4 (1999): 23–49.

Somerville, Kristine. "Stage Pictures: Jo Mielziner and the Art of Set Design." *Missouri Review* 42, no. 3 (2019): 55–65.

Spellman, Francis. "Communism Is Un-American." *American Magazine*, July 1946, 26–28.

Sugden, Robert P. *Care and Handling of Art Objects*. New York: Metropolitan Museum of Art, 1946.

Tablet. "Civitas Dei." January 28, 1959, 8.

Tablet. "Pieta to Spend Winter in Vatican Pavilion." October 15, 1964.

Tablet. "Set Dedication of Vatican Pavilion at World's Fair." October 10, 1963.

Talese, Gay. *The Kingdom and the Power*. New York: World, 1969.

Time. "Americans at Brussels." June 16, 1958.

Time. "The Beleaguered Man." April 4, 1955.

Time. "The Year in Books." Books. December 18, 1950.

Time. "Roman Catholics: The Pastor-Executive." May 15, 1964.

Tirella, Joseph. *Tomorrow-Land: The 1964–65 World's Fair and the Transformation of America*. Guilford, CT: Lyons, 2013.

Todd, Jesse T. "Imagining the Future of American Religion at the New York World's Fair, 1939–40." PhD diss., Columbia University, 1996.

Tolnay, Charles de. *The Youth of Michelangelo*. Princeton, NJ: Princeton University Press, 1969.

Vasari, Giorgio. 1991. *The Lives of the Artists*. Translated by Julia Conway Bondanella and Peter Bondanella. New York: Oxford University Press.

Vatican Pavilion, New York World's Fair 1964–1965: A Chronicle. New York: New York World's Fair, 1966.

Wallace, William E. "An Impossible Task." In *Making and Moving Sculpture in Early Modern Italy*, edited by Kelley Helmstutler Di Dio, 3–58. Burlington, VT: Ashgate, 2015.

Warner, Dennis. *The Last Confucian: Vietnam, South-East Asia, and the West*. New York: Macmillan, 1963.

Weinraub, Bernard. "A City Transfixed." In Rosenthal and Gelb, *Pope's Journey*, 52–54.

WGBH Radio. "Vietnam: A Television History; America's Mandarin (1954–1963); Interview with Jack Keegan, 1981." May 12. https://openvault.wgbh.org/catalog/V_96208C5C09F7462084BDDA91206A70C5.

Whitman, Alden. *Come to Judgment*. New York: Viking, 1980.

Wicker, Tom. "A Talk with the President." In Rosenthal and Gelb, *Pope's Journey*, 14–19.

Wilkinson, Ed. "Readers Recall World's Fair Memories." *Tablet*, September 3, 2014. https://thetablet.org/readers-recall-worlds-fair-memories/.

Wintoff, Naj. "On the Scene: Duval Opens 'If Not Now, When' Exhibit at Keene Arts." *Lake Placid (NY) News*, August 12, 2021.

WNYC (New York Public Radio). "Flushing Meadows Corona Park, Queens Dedication Ceremonies." June 2, 1967. wnyc.org/story/flushing-meadows-corona-park-queens-dedication-ceremonies/.

Zhuang, Justin. "Design History 101: Quietly Beautiful Work by the Illustrator Who Drew the Four Seasons Logo." April 9, 2015. eyeondesign.aiga.org/design-history-101-quietly-beautiful-work-by-illustgrator-who-drew-four-seasons-logo.

Zide, Larry. "Inzide Audio." *Audio*, April 36, 1965.

Index

Figures are indicated by "f" following page numbers.

www.ingramcontent.com/pod-product-compliance
Lightning Source LLC
LaVergne TN
LVHW091644100826
845152LV00010B/162/J

* 9 7 8 1 5 0 1 7 7 6 9 0 8 *